Late Ordovician brachiopods from West-central Alaska: systematics, ecology and palaeobiogeography

by

Christian M. Ø. Rasmussen, David A. T. Harper
and Robert B. Blodgett

Acknowledgement

Financial support for the publication of this issue of Fossils and Strata was provided by the Lethaia Foundation

Contents

Late Ordovician brachiopods from West-central Alaska: systematics, ecology and palaeobiogeography

CHRISTIAN M. Ø. RASMUSSEN, DAVID A. T. HARPER and ROBERT B. BLODGETT

Rasmussen, C. M. Ø., Harper, D. A. T. & Blodgett, R. B. 2012. Late Ordovician brachiopods from West-central Alaska: systematics, ecology and palaeobiogeography. *Fossils and Strata*, No. 58, pp. 1–103. ISSN 0024–1164.

A silicified brachiopod fauna from the White Mountain area, west-central Alaska is catalogued and interpreted within a palaeoecological and biogeographical context. This area is situated within the Nixon Fork Subterrane of the Farewell Terrane; its origin and timing of final docking with Laurentia has been much debated. The current study adds new faunal data to the debate. The fauna was collected from three localities; a main locality in the upper Darriwilian – lower Sandbian and two additional middle–upper Katian localities. All three localities contain a predominantly deep-water autochthonous fauna that was mixed with an allochthonous fauna as a result of down-slope movement of turbidity currents at the shelf-break to slope transition. Deposition of the autochthonous fauna was within Benthic Assemblage zones 4–5, possibly shallower in the Katian localities. The three localities comprise a very diverse fauna consisting of nearly 100 taxa. Of these the genera *Callositella cheeneetnukensis*, *Duolobella sandiae*, *Palaeowingella farewellensis* and *Transridgeia costata* are new. In addition nine new species are described: *Anisopleurella tricostata*, *Christiania aseptata*, *Craspedelia potterella*, *Gelidorthis perisiberiaensis*, *Leptaena (Septomena) alaskensis*, *Oanduporella kuskokwimensis*, *Ptychoglyptus alaensis*, *Sowerbyella (Rugosowerbyella) praecursor* and *Sowerbyella (S.) rectangularis*. Furthermore, *Leptellina pulchra* and *Strophomena planobesa* are revised and assigned to *Anoptambonites* and *Tetraphalerella*, respectively. The large number of new taxa indicates an endemic fauna that has affinities with the cosmopolitan deep-water faunas of this time interval. However, the fauna demonstrates great similarity with those from the Eastern Klamath Terrane, northern California and the Jones Ridge area on the Alaska – Yukon border, as well as faunas of Eastern USA. As the Laurentian affinities are all based on deep-water taxa, it may not indicate particularly close faunal links between the Farewell Terrane and Laurentia. More noteworthy, several taxa indicate faunal exchange, in some cases at the species level, between peri-Laurentian terranes such as the Midland Valley Terrane and the palaeocontinents of Siberia and even Baltica. Notably the link to Siberia is strengthened by two cluster analyses conducted for the studied localities, indicating that the Farewell Terrane probably is derived from Siberia but by the Late Ordovician was in a sufficiently isolated position to develop a whole suite of endemic taxa.

□ *Late Ordovician, Farewell Terrane, peri-Siberia, brachiopods, palaeoecology, biogeography*.

Christian M. Ø. Rasmussen [christian@snm.ku.dk], Center for Macroecology, Evolution and Climate and Nordic Center for Earth Evolution (NordCEE), Natural History Museum of Denmark, University of Copenhagen, Øster Voldgade 5–7, DK-1350 Copenhagen K, Denmark; David A. T. Harper [david.harper@durham.ac.uk], Department of Earth Sciences, Durham University, Durham DH1 3LE, UK; Robert B. Blodgett [robertbblodgett@yahoo.com], 2821 Kingfisher Drive, Anchorage, Alaska 99502, USA; manuscript received on 17/08/2010; manuscript accepted on 12/10/2010.

Introduction

Alaska consists of numerous terranes that have accreted onto the North American craton during the Mesozoic. The origin of these terranes has been a continuing subject of discussion (e.g. Blodgett & Boucot 1999; Dumoulin *et al.* 2000; Blodgett *et al.* 2002).

The Late Ordovician was an interval of rapid global change. Continents drifted rapidly towards the Equator where Laurentia, the largest of the equatorial palaeocontinents, was located (Torsvik *et al.* 1996; Fortey & Cocks 2003; Cocks & Torsvik 2011). The same interval was dominated by what was probably a Phanerozoic sea level maximum (Hallam 1984; Nielsen 2004). As a result continents were flooded by extensive epicontinental seas; a phenomenon not known on the Earth today. This promoted ideal living conditions for the marine benthos, such as the brachiopods, which as a result rapidly increased globally in numbers of taxa. Thus, the combination of the palaeogeographic configuration, the high sea level and the resultant high γ-diversity was of great importance with respect to the endemism of rather isolated continents like Baltica and a series of smaller terranes like those that amalgamated with Laurentia during the Mesozoic to form present-day Alaska.

10.1111/j.1502-3931.2011.00298.x

The current study presents a systematic overview of the brachiopods from one of these terranes, the Farewell Terrane. This study is based on material collected during the late 1970s, from which only lists of genera have been published to date (Potter *et al.* 1980, 1988; Potter 1984, 1990c, 1991; Potter & Blodgett 1992; Potter & Boucot 1992). The fauna consists of late Mid to Late Ordovician (late Darriwilian – late Katian) brachiopods from the White Mountain area in west-central Alaska (McGrath A-4 and A-5 quadrangles, Upper Kuskokwim River region). During the past three decades, opinions as to where the Farewell Terrane was positioned in the Late Ordovician have shifted from a peri-Laurentian position to, most recently, an integral part of Siberia or Kolyma (Potter 1984; Blodgett & Clough 1985; Rohr & Blodgett 1985; Potter *et al.* 1988; Adrain *et al.* 1995; Blodgett & Boucot 1999; Dumoulin *et al.* 2000; Blodgett *et al.* 2002, 2010; Rigby *et al.* 2009; Cocks & Torsvik 2011). The current study is in line with the most recent literature, as it indicates a close link to Baltica and Siberia. However, the closest faunal similarities are with the Eastern Klamath Terrane, which was positioned far off the western Laurentian margin close to the

Alexander Terrane of present day SE Alaska, which, in turn, is regarded closely related to the Farewell Terrane (Potter 1990a; Blodgett *et al.* 2010).

Geological setting

Nearly all of Alaska is made up of a series of terranes that have accreted to the North American continental margin in the Mesozoic. The only exception is the triangular area in east-central Alaska on the Alaska – Yukon border (Charley River A-1 quadrangle), bounded roughly on its NW and SW sides by the Porcupine and Yukon Rivers, respectively, which includes the Jones Ridge area of the Nation Arch and the Porcupine Plateau area to the north (Blodgett *et al.* 2002, 2010; Brabb 1967; Ross & Dutro 1966; Fig. 1).

The Farewell Terrane is composed of the Nixon Fork, Dillinger and Mystic subterranes (Blodgett *et al.* 2002, 2010; Decker *et al.* 1994; Dumoulin *et al.* 1998; Fig. 1). The White Mountain area is situated within the Nixon Fork Subterrane. Ordovician brachiopods from the Nixon Fork Subterrane are reported from three sites: Telsitna Ridge in the northern Kuskokwim

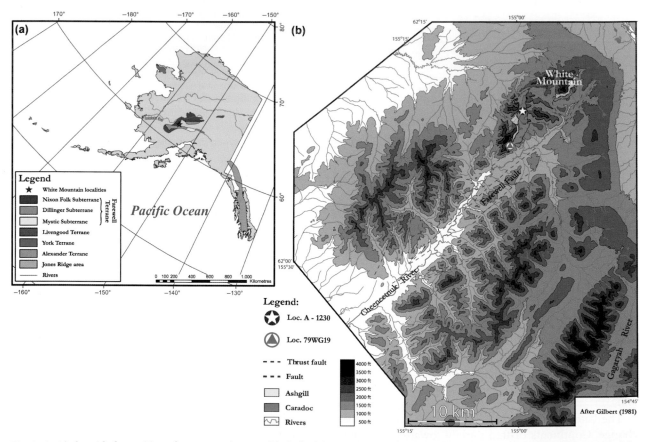

Fig. 1. A, Alaska with the position of terranes and areas with Ordovician strata. Note the geographic position of the Farewell Terrane and its subdivisions. Based on Blodgett *et al.* (2002) and Brabb (1967). B, Simplified map of the White Mountain area, McGrath A-4 and A-5 quadrangles, west-central Alaska. Based on the geological map by Gilbert (1981) for the Cheeneetnuk River area. For explanation of colours, see legends. Note that locality 79WG126 is not shown, but should be positioned in proximity to locality A-1230.

Mountains, Lone Mountain and White Mountain (Potter & Blodgett 1992). The Telsitna Ridge locality differs from the others in being a shallow-water carbonate succession in contrast to the deep-water carbonates and siliciclastics constituting the two other localities (Measures *et al.* 1992; Rohr *et al.* 1992; see also Fig. 2). The geographic position of the Farewell Terrane and its subdivisions are shown in Figure 1 together with the positions of two of the studied localities in the White Mountain area; the third locality lies close proximity to locality A-1230, but has not precisely been located.

Study area

The material under study was collected by W. G. Gilbert in 1978 and A. W. Potter, W. G. Gilbert and R. B. Blodgett in 1979 from the White Mountain area, west-central Alaska (McGrath A-4 and A-5 quadrangles).

Three localities have been investigated. In previous literature (e.g. Rigby *et al.* 1988; Potter & Boucot 1992; Candela 2006b) locality A-1230 and locality 79WG19 were designated units uOI and uOII, respectively. These units are based on the geological map published by Gilbert (1981) of the Cheeneetnuk River area (which includes White Mountain), indicating the two different Upper Ordovician limestone units in the area (Fig. 1). Locality 79WG126 has not previously been mentioned in the literature.

It should be noted that the University of Alaska Museum locality number A-1230 corresponds to Wyatt Gilbert's field locality 78WG162, but has subsequently been changed by the University of Alaska Museum when the original collection was accessioned into their collections. As A-1230 has been used in previous literature, we will continue this use.

Localities

Locality A-1230. – This is the main locality; it is approximately 15 metres in diameter positioned on a northwest trending slope at an altitude of about 670 metres near the lowest part of a saddle (62°09,4′N, 154°59,6′W; Fig. 1). Six loose limestone blocks were collected at this locality (Rigby *et al.* 1988). These are associated with a siltstone-shale succession with interbedded limestone. The limestone beds were deposited below storm-wave base by debris flows and turbidity currents beyond the shelf margin break or on a distally steepened ramp (Potter *et al.* 1988; Rigby *et al.* 1988). Locality A-1230 is part of the uOI mapping unit of Gilbert (1981).

Locality 79WG19. – This locality is situated on the crest of a SE trending spur of hill 3418 at an altitude of about 625 metres, McGrath A-5 quadrangle (62°07,3′N, 155°01,2W; Fig. 1). The blocks were derived from a chert-argillite. Locality 79WG19 is situated in the uOII mapping unit of Gilbert (1981).

Locality 79WG126. – The position of this locality is not precisely known. The Alaska State Geological Survey (formally known as the Alaska Division of Geological & Geophysical Surveys) does not have copies of Wyatt Gilbert's field notes or maps from the late 1970s. One of us (Blodgett) has several old photostatic (xerox) copies of some of Gilbert's 1979 field maps showing the localities 79WG120, 121, and 129 in the SW corner of the McGrath A-4 quadrangle (close to locality A-1230). We suspect 79WG126 was near here, probably close to locality A-1230, and situated in the uOII unit of Gilbert (1981). The material was derived from a chert-argillite.

Extraction methods

Together with other fossil groups, such as corals, gastropods and sponges (Rohr & Blodgett 1985; Rigby *et al.* 1988), the brachiopod shells have been silicified. This silicification has afforded excellent preservation. The specimens, however, are very fragile and easy to break. The extraction process (see below) may further have contributed to the fragmentation of the allochthonous parts of the fauna.

Brachiopod shells were retrieved from the limestone matrix by the use of hydrochloric acid and selected from the insoluble residues. Potter undertook the painstaking task of gluing the many brachiopod fragments back together. Because of the extremely fragile state of the valves, we have not tried to remove the glue. Thus, the glue is sometimes visible in the SEM images presented below. This was impossible to avoid. The excess of glue on the specimens, in some cases, made it difficult to classify for example dalmanelloid brachiopods, which are usually separated from the Orthoidea by the presence of exopunctae.

Most specimens have been photographed using a FEI Inspect SEM microscope. However, the larger specimens were photographed using a Canon EOS D20 digital camera with a 100 mm macro zoom lens.

Age of the fauna

Sainsbury (1965), who carried out the initial reconnaissance in the White Mountain area, reported Middle Ordovician rocks based on the occurrence brachiopods such as *Christiania* Hall & Clarke, 1892. Despite the recognition of the Ordovician strata in that area nearly 50 years ago, to date no formal lithostratigraphic names have been established for these strata. Thus, it is not possible to assign the fauna of the three localities studied here to formally named units. However, estimates of the age of the rock units that comprise localities A-1230, 79WG19 and 79WG126, have been proposed.

As mentioned above, six limestone blocks were collected from loose blocks at locality A-1230 (Rigby *et al.* 1988). Their ages were determined using either conodont (determined by Stig M. Bergström, unpublished data) or brachiopod faunas (Potter 1984; Potter & Blodgett 1992). According to Rigby *et al.* (1988), the age ranges were as follows:

Conodonts from block 1 were of late Sandbian – early Katian age; block 2 was Sandbian or possibly younger; block 3 was Sandbian although it consists of a combination of older conodonts with a younger macrofauna; blocks 4, 5 and 6 were late Darriwilian – early Katian in age on the basis of a mixed assemblage of macrofossils.

The present study is based mainly on material from blocks 4–6, which constrains the age to the late Darriwilian – early Sandbian according to the occurrences of the following taxa: *Anoptambonites grayae* Davidson, 1883; *Anoptambonites pulchra* (Cooper, 1956), *Desmorthis*? Ulrich & Cooper, 1936; *Diorthelasma parvum* Cooper, 1956; *Doleroides* Cooper, 1930; *Leptellina occidentalis* Ulrich & Cooper, 1936; *Leptellina tennesseensis* Ulrich & Cooper, 1936; *Lepteloidea leptelloides* Bekker, 1921, *Perimecocoelia semicostata* Cooper, 1956; *Scaphorthis virginiensis* Cooper, 1956; *Taphrorthis immatura* Williams & Curry, 1985; *Tetraphalerella planobesa* (Cooper, 1956) and *Xenambonites revelatus* Williams, 1962. Moreover, the occurrence of *Didymelasma* Cooper, 1956; *Sowerbyites* Teichert, 1937; *Oanduporella* Hints, 1975 and *Leptaena (Septomena)* Rõõmusoks, 1989 are confined to the lower Sandbian – lower Katian, making it highly unlikely that these blocks are younger. However, *Sowerbyella (Rugosowerbyella)* Mitchell, 1977 has not previously been recorded from beds as old as lower Sandbian. Thus, this may be the oldest occurrence of this subgenus. The same is the case with *Tetraphalerella*, which previously was believed only to occur in Katian strata. However, in the current study, the early Sandbian *Strophomena planobesa* Cooper, 1956 has been transferred to *Tetraphalerella*, thus extending the generic range of the latter to the lower Sandbian.

Locality 79WG19 was previously reported to contain early Katian–Hirnantian brachiopods (Potter 1984; Potter & Blodgett 1992). Current revision of the brachiopods from this locality indicates an early to late Katian age (Pusgillian–Rawtheyan). The lower boundary is constrained by the occurrence of *Dicoelosia jonesridgensis* Ross & Dutro, 1966; *Diochthofera* aff. *conspicua* Potter, 1990a, *Catazyga* Hall in Hall & Clarke, 1893, *Kassinella (Trimurellina*?) Mitchell, 1977; whereas the upper boundary is constrained by the occurrence of *Anoptambonites, Catazyga, Cyclospira orbus* Cocks & Modzalevskaya, 1997 and *Grammoplecia*? Wright & Jaanusson, 1993; *Brevicamera*? Cooper, 1956, occurs at this locality although it is previously only known from the upper Darriwilian. This extends considerably its stratigraphic range into the Katian. Alternatively, this may be a new, closely related genus.

The fauna from the blocks at Locality 79WG126 is constrained to the latest Katian (Rawtheyan), based on the occurrence of the early virgianid *Galeatellina* Sapelnikov & Rukavishnikova, 1976.

Based on the new, better constrained age estimates, the White Mountain fauna is tentatively compared with those of adjacent palaeoplates and terranes in Figure 2.

Although the present study has constrained the age range of much of the fauna, the data matrices used for Figures 5 and 6 are nonetheless based on the broader age determinations of previous studies, primarily to obtain a larger data set. Secondly, in the case of Figure 5, it has been easier to confine the cluster analysis to the Caradoc Series of the British terminology. Therefore, the data matrix from locality A-1230 includes taxa from the whole of the Caradoc (Sandbian–lower Katian) (Fig. 5), whereas the matrix for localities 79WG19 and 79WG126 (Fig. 6) includes taxa of early–mid Ashgill (mid–late Katian) age.

The brachiopod fauna

2461 specimens were available for study, which were referred to 96 species within probably as many as 84 different genera. Of these, some 21 taxa have not been identified to the genus or – in some cases – even family level. This is due to the fragmented state of the material. Often only the cardinalia are preserved. Thus, only in the cases of taxa with very characteristic cardinalia could the fragmented material be assigned to a genus. The fact that cardinalia are often preserved in the material, allowed to identify 98% of the valves: 50% are ventral valves, 44% are the dorsal counterpart and 4% represents articulated valves. For a small-shelled, silicified fauna deposited in a turbidite environment, the percentage of preserved dorsal valves is apparently high. This is explained by the high percentage of plectambonitoid taxa, 27%, that constitute the fauna. In terms of actual valves, the Plectambonitoidea

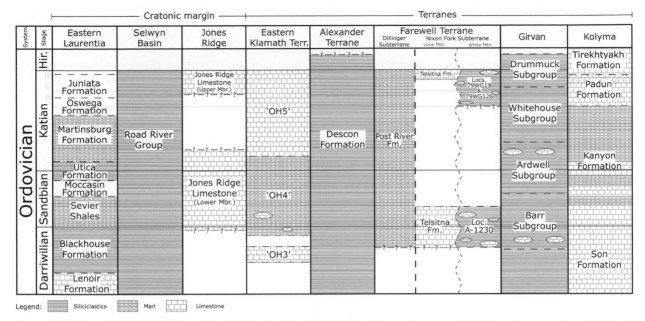

Fig. 2. Correlation of Ordovician lithostratigraphical units with adjacent palaeoplates and terranes which share important faunal links with the Farewell Terrane. The Farewell Terrane has been divided into the deep-water Dillinger Subterrane and the mixed facies of the Nixon Fork Subterrane. The latter is further sub-divided into the shallow-water Lone Mountain facies, as well as the deep-water White Mountain facies, that are the topic of the current study. Question marks denote boundaries with poor biostratigraphical control. P, Pusgillian; C, Cautleyan; R, Rawtheyan. The following sources were used to compile the chart: *Eastern Laurentia* (Tennessee and Virginia): Cooper (1956), Ettensohn (2008); *Selwyn Basin*: Miall (2008); *Jones Ridge*: Brabb (1967), Churkin & Carter (1970), Ross & Dutro (1966); *Eastern Klamath Terrane*: Potter (1990a); *Alexander Terrane*: Blodgett *et al.* (2010), Churkin & Carter (1970); *Farewell Terrane*: Blodgett *et al.* (2002), Dumoulin *et al.* (1998), Measures *et al.* (1992), Rohr *et al.* (1992), this study; *Girvan* (Midland Valley Terrane): Harper (2006), Ingham (2000); *Kolyma*: Oradovskaya (1988), Zhang and Barnes (2007).

represent 30% of the material. Plectambonitoids have much larger and thicker dorsal valves than their ventral valves and thus have not fractured as easily during post-mortem transport. Furthermore, 10% of the fauna is assigned to the Strophomenoidea, which have similarly large and thick dorsal valves (i.e. *Leptaena* Dalman, 1828).

The fauna is dominated by the Orthida. Together the Plectorthiodea, Orthoidea, Dalmanelloidea, Enteletoidea and a group of unassigned orthidinids comprise some 50% of the taxa. With regard to the number of specimens, the Orthida contain 37% of the material. In terms of specimens, the plectorthoids dominate the Orthida, with 24% of the taxa. The remaining superfamilies that are represented in the material are the Anazygoidea, Atrypoidea, Camerelloidea, Clitambonitoidea, Discinoidea, Pentameroidea, Protozygoidea, Skenidioidea and Triplesioidea. Together these comprise the remaining 13% of the taxa.

At the species level the fauna is dominated by *Ptychopleurella uniplicata* Cooper, 1956 with 298 shells collected. Other dominant species are *Scaphorthis virginiensis* Cooper, 1956; *Taphrorthis immatura* Williams & Curry, 1985; *Christiania aseptata* n. sp., *Anoptambonites grayae*, *Sowerbyella* (*S.*) *rectangularis* n. sp., *Xenambonites revelatus*, *Skenidioides multifarius* Potter, 1990b and *Phragmorthis buttsi* Cooper, 1956.

These taxa are generally deep-water forms (Fig. 3). The ecological aspects of the fauna are discussed further in the next section.

The following genera listed from the two White Mountain localities by Potter (1984, 1990c, 1991), Potter & Blodgett (1992), Potter *et al.* (1988), Potter & Boucot (1992) have not been confirmed by the current study: *Paralenorthis* Havlíček & Branisa 1980 from locality A-1230 and *Bimuria* Ulrich & Cooper (1942), *Eospirigerina* Boucot & Johnson (1967), *Eripnifera* Potter (1990a) and *Tcherskidium* Nikolaev & Sapelnikov (1969) from locality 79WG19. *Tcherskidium*, however, is with certainty reported from other localities within the Nixon Folk Subterrane (Blodgett *et al.* 2002, 2010).

Palaeoecological aspects

Previous studies indicated that the material comprises a shallow-water allochthonous fauna that has been transported downslope by turbidity currents and mixed with an autochthonous deeper-water fauna (Potter 1984; Potter *et al.* 1988; Potter & Blodgett 1992). A depth-range within Benthic Assemblage Zone (BA) 4 of Boucot (1975) was proposed by Potter (1990a) and Potter & Boucot (1992).

Normal wavebase	Photic zone ~ 200 m	Low tide		High tide		
Storm wavebase			Subtidal zone	Intertidal zone	Supratidal zone	
BA 6	BA 5	BA 4	BA 3	BA 2	BA 1	
Abyssal plain	Continental slope	Continental shelf				

Count	Genus
2	*Stenocamara*
4	*Acrosaccus*
6	*Doleroides*
33	*Dinorthis*
14	*Kassinella (Trimurellina?)*
2	*Anazyga*
25	*Sow. (Rugosowerbyella)*
5	*Zygospira*
1	*Desmorthis*
1	*Didymelasma?*
2	*Galeatellina*
13	*Leptaena (Septomena)*
1	*Sowerbyites*
23	*Tetraphalerella*
79	*Gelidorthis*
98/ 1	*Sowerbyella (S.)*
1	*Atelelasma*
3	*Austinella*
18	*Catazyga*
19/ 1	*Hesperorthis*
7	*Perimecocoelia*
37	*Plectorthis*
3	*Eoanastrophia*
16	*Rhactorthis*
1	*Grammoplecia*
4	*Camerella*
2	*Diochtofera*
14	*Leptelloidea*
4	*Replicoskenidioides*
4	*Salopina*
25	*Leptaena (L.)*
1	*Schizostrophina*
4	*Epitomyonia*
200/ 4	*Anoptambonites*
33	*Ptychoglyptus*
14	*Leptellina*
7	*Paucicrura*
200/ 16	*Ptychopleurella*
1	*Brevilamnulella*
11	*Diambonia*
214	*Taphrorthis*
13	*Triplesia*
159/ 1	*Christiania*
91/ 7	*Skenidioides*
3/ 49	*Cyclospira*
2	*Brevicamera*
58/ 1	*Craspedelia*
105	*Xenambonites*
11	*Laticrura*
49	*Oanduporella*
39	*Bimuria*
92/ 4	*Phragmorthis*
112	*Scaphorthis*
16	*Orthorhynchuloides*
70	*Diorthelasma*
76	*Anisopleurella*
13	*Dicoelosia*
15	*Eoplectodonta (E.)*
51/ 5	*Glyptorthis*

Fig. 3. Depth distribution of the genera found in the current material positioned according to the Benthic Assemblage zones of Boucot (1975). The ranges of the individual genera are based on the literature. Blue colour indicates locality A-1230, yellow indicates localities 79WG19 and 79WG126. If a genus is indicated by both colours, it is present at both the older and younger localities.

The current investigation supports this view, with the bulk of the taxa being assigned to within the BA 3–5 zones (Fig. 3). But there is also indication of more lengthy transport. Examples are the acrotretoid *Acrosaccus* cf. *shuleri* Willard, 1928 and *Doleroides*. The latter genus, which is represented by two species in the current material, occupied an intertidal to restricted shallow, subtidal environment in eastern North America (Patzkowsky 1995). This confirms the occurrence of the genus within shallow-water platform carbonates of the northern part of the Nixon Fork Subterrane (Rohr *et al.* 1992). At the other end of the spectrum, genera that usually belong to BA 5–6, like *Glyptorthis*, also occur within the studied material.

Even though most of the material is heavily fragmented, the silification has also allowed for some very well preserved specimens. For example, borings are seen in some valves, indicating that the shells may have been subject to predation. Furthermore, some specimens have been found with minute incrusting juvenile specimens of *Ptychopleurella uniplicata*, *Skenidioides multifarius* and *Taphrorthis immatura* on larger orthide or strophomenoid shells (Fig. 4). This suggests that the above-mentioned three species have not been transported far, and thus probably belong to the BA 4–5 assemblage. This is further supported by their frequency, for instance *P. uniplicata* being the most abundant species in the material with nearly 300 valves. Other dominant taxa are *Scaphorthis virginiensis* and *Taphrorthis immatura* together with *Christiania aseptata*, *Anoptambonites grayae*, *Sowerbyella* (*S.*) *rectangularis*, *Xenambonites revelatus*, *Skenidioides multifarius* and *Phragmorthis buttsi*. As already mentioned, all of these taxa are known to favour a shelf margin to upper slope setting. This is compatible with the view that this material primarily reflects a deeper-

water autochthonous fauna mixed with allochthonous elements derived from various settings seaward from the nearshore environment. This was probably the case for both the Sandbian (loc. A-1230) and Katian (locs. 79WG19 and 79WG126) faunas, though the latter may have lived in slightly shallower-water environments based on the analysis presented in Fig. 3. In addition, the virgianid genera *Brevilamnulella* and *Galeatellina* occur at localities 79WG19 and 79WG126. Both genera are often associated with deep-water carbonate mound settings in the late Katian, suggesting that carbonate mounds may have been present somewhere on the slopes of the Nixon Fork Subterrane at this time.

When compared with cratonic Laurentia, this depth assignment compares well with the *Paucicrura*–plectambonitacean biofacies of the Oranda and Edinburg formations of Virginia and Tennessee (Patzkowsky 1995). The Alaskan material has eight of the 14 genera listed for this biofacies, twice as many as for either of the two shallow-water biofacies interpreted by Patzkowsky (1995). The *Paucicrura*–plectambonitacean biofacies was interpreted as being deposited offshore, below storm wave base, near the shelf margin break (Patzkowsky 1995). This further supports the view that the Laurentian biofacies had a broad geographic significance (Potter & Boucot 1992; Patzkowsky 1995), emphasizing also the strong affinities of the Farewell Terrane with cratonic Laurentia (see next section).

Palaeogeographical significance

As mentioned in the introduction, previous studies on the Farewell Terrane have argued for quite different views on its origination and, thus, its geographic

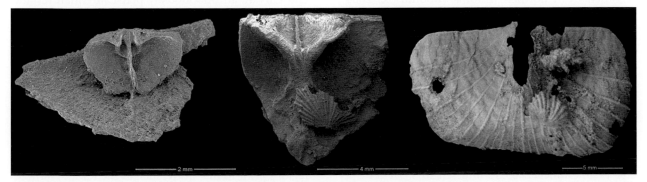

Fig. 4. Images demonstrating some palaeoecological aspects of the fauna. From left: a minute juvenile dorsal specimen of *Skenidioides multifarius* encrusting a silicified fragment (MGUH 29679), center: a juvenile dorsal valve of *Taphrorthis immatura* encrusting an adult specimen of the same species and right: a conjoined juvenile specimen of *Ptychopleurella uniplicata* encrusting a large ventral valve of Strophomenoidea gen. et sp. indet. 2 (MGUH 29464). Whereas the large strophomenoid may have been subjected to a lenthy transport from the shelf to upper slope settings, the minute juvenile encrusters indicate that they are probably part of an autochthonous fauna.

position in the Late Ordovician. According to Potter *et al.* (1988) the Farewell Terrane shows the closest links at the species level with, in particular, the Jones Ridge area, which is believed to have been an integral part of the North American craton in the Late Ordovician. Further, close faunal links are shared with the Arctic Alaska Terrane (which includes the Brooks Range), the Yukon Territory, the Alexander Terrane of south-eastern Alaska and the Klamath Mountains (Eastern Klamath Terrane) in northern California. These views were further supported by analysis of comparative sponge data (Rigby *et al.* 1988). Potter *et al.* (1988) further found the fauna to show more exotic links with Chukotka in Northeast Russia, Kazakhstan and South China. In addition links to Baltica via the Taimyr Peninsula in northern Siberia have been proposed by Cocks & Modzalevskaya (1997).

Within the last decade notably the link to Siberia has gained considerable weight. Blodgett *et al.* (2002) found strong faunal affinities to Siberia within Middle Cambrian to Middle Devonian faunas of the Nixon Fork Subterrane. Since then several studies on various faunal elements have confirmed this view (Holloway 2004; Rigby *et al.* 2005, 2008, 2009). Therefore, both the Farewell and Alexander terranes (as well as other Alaskan terranes) are now believed to have been positioned closer to Siberia than Laurentia (Blodgett *et al.* 2010).

Contrary to previous faunal lists published on the White Mountain fauna, the links to Siberia are now also supported by *Cyclospira elegantula* Rozman, 1964 from the Sandbian–lower Katian of Kolyma (Rozman 1964) and the Chinghiz Terrane (Klenina *et al.* 1984) and from the upper Katian, *Cyclospira orbus*, hitherto known only from the upper Katian of central Taimyr (Cocks & Modzalevskaya 1997) and *Eoanastrophia?* The links to Baltica are also considerably strengthened by the presence of the leptaenid subgenus *Septomena*, which, prior to the current study, has only been recorded from northern Estonia (Rõõmusoks 2004), *Ptychoglyptus pauciradiatus* Reed, 1932, is known from the Trondheim area, Norway (Reed 1932), although this probably was a Laurentian marginal terrane during the Late Ordovician, *Leptelloidea* Jones, 1928; known from Estonia and the Southern Shan states (Cooper 1956; Cocks & Rong 1989) and the dalmanelloid orthid *Oanduporella*, previously thought to be a late Sandbian – early Katian Baltic endemic, but now known to have a much wider geographic distribution (Hints 1975; Parkes 1992; Suárez-Soruco 1992; Benedetto 1995; Candela 2003; Rasmussen 2011). A new species of *Oanduporella* occurs in this material, suggesting that the centre of origin of the genus was not, as previously thought, in Baltica where the youngest occurrence is now found. In addition, one specimen, which

may be assigned to *Sampo* Öpik, 1933, and a fragment questionably assigned to *Grammoplecia*, could further indicate this Baltic biogeographic affinity.

Nonetheless, the closest links of the current fauna are undoubtedly with the Eastern Klamath Terrane in northern California with which it, despite of the fact that probably the highest eustatic sea-level stand in the Phanerozoic was during the Sandbian, still maintained links at the species level (Hallam 1984, 1992; Nielsen 2004). The Eastern Klamath Terrane was situated somewhere off the North American craton in the Late Ordovician. Recent studies suggest that it was likely farther away from Laurentia than previously thought and was, much like the Farewell Terrane, positioned close to Siberia at least until the latest Devonian (Rigby *et al.* 2005). Regardless of the timing of final docking with Laurentia, the Eastern Klamath Terrane was probably, like the Alexander and Farewell terranes, a forerunner of the many terranes, which accreted to Laurentia during the Palaeozoic and Mesozoic. But compared to the fauna from the Eastern Klamath Terrane (Potter & Boucot 1971; Potter 1990a,b,c, 1991) the current study demonstrates that the brachiopods from the Nixon Fork Subterrane clearly maintained strong links to Siberia and Baltica, though they may appear more subtle than the many shared species with Laurentia and also Girvan. However, these affinities are mostly based on deep-water taxa. Arguably, the links to Siberia and Baltica found in the current study are taxa originating in typically more shallow water facies. This would favour a more peri-Siberian position of the studied material, as shallow water taxa usually have a narrow geographical distribution.

Some of the most abundant species within the White Mountain material are *Anoptambonites grayae*, *Scaphorthis virginiensis*, *Taphrorthis immatura* and *Xenambonites revelatus*. These species have only previously been found in cratonic Laurentia (the Jones Ridge and the 'The Great American Carbonate Bank') and Girvan, an undoubtedly peri-Laurentian terrane. Thus, they link the Nixon Fork Subterrane to these regions. However, as they are deep-water taxa they may also signal that these tropical terranes probably shared similar deep-water faunas. A genus such as *Gelidorthis* is a more exotic element within the fauna as it, prior to this study, only was known from siliciclastic settings in peri-Gondwana (Botquelen & Mélou 2007; Villas 1992). Similar to *Gelidorthis*, genera such as *Galeatellina*, *Eoanastrophia* and *Schizostrophina*, to name a few, are all known exclusively from exotic regions prior to this study, such as the Kazakh terranes and South China. Thus, the fauna also contains a clear cosmopolitan component.

However, the endemic character of the White Mountain fauna is strongly emphasized by its representatives of the strophomenoid subfamily, the Furcitellinae, and a handful of new genera that all indicate that the Farewell Terrane must have been sufficiently isolated by the Late Ordovician to develop these genera. Several more taxa are probably new species, if not genera. However, due to the fragmented state of much of the material, these are left under open nomenclature. Albeit, despite the overwhelming amount of evidence of comparable strata and fauna within especially the Silurian and Devonian successions of the Alexander Terrane and Siberia (Blodgett *et al.* 2010), the endemic signal of the White Mountain fauna seems too strong to position the Farewell Terrane as an integral part of Siberia during the Late Ordovician.

There is no doubt, that previous assessments of the biogeographical affinities of the Farewell Terrane, have been strongly influenced by the apparent strong Laurentian signal due to the many shared species. However, this has not previously been statistically tested and thus, in our view, too much emphasis has been given to near cosmopolitan deep-water species. To monitor the above discussed faunal affinities statistically, cluster analyses were performed for the localities studied. These are presented in Figures 5 and 6. The cluster matrices differentiate between facies across the larger palaeocontinents. This was undertaken to separate the cosmopolitan faunas from the more endemic, shallow-water faunas. Thus, Baltica is separated into shallow and deeper-water clusters. The shallow-water cluster is based on East Baltic shallow-water carbonates and Öland, Sweden. These occur approximately within the BA 1 to upper BA 4 zone. The deeper-water cluster is based on the Oslo Region, Norway, the Holy Cross Mountains in southern Poland, Västergötland, central Sweden, and the Scania and Siljan areas, southern and central Sweden, respectively. These are mostly siliciclastics except for the Dalby, Boda and Kullsberg limestones in Siljan, the latter two being carbonate mud mounds with core and flank facies thought to have been deposited within the BA 4–5 zones (Cocks 2005; Rasmussen *et al.* 2010). With regard to Laurentia, the faunas from the following formations are regarded as deep-water faunas: Lower Sevier Shale, Edinburg Formation, Effna–Rich Valley Formation, Chatham Hill Formation, Lower Bromide Formation, Arline Formation, Decorah Shale, Martinsburg Formation, Oranda Formation, Little Oak Formation. In addition, the Pyle Mountain Argillite, Maine, the Sansom Formation, Newfoundland, part of the White Head Formation, Quebec, most of the Vaureal Formation, Anticosti Island, part of the Børglum River Formation, eastern North Greenland and the Advance Formation of British Columbia are assigned to deeper-water faunas of Laurentia. In North China the

Pingliang Formation is associated with deep water and finally, in South China, the Huangnekang, Miaopo, Pagoda and Tangtou formations were deposited in deep-water environments.

The cluster analysis is based on a very large matrix built using relevant literature. More specific information as to the compilation method and possible biases of the database can be found in Rasmussen and Harper (2011a, b). The Sandbian – early Katian data matrix is based on 342 genera distributed across 26 locations and the mid–late Katian matrix on 289 genera in 24 locations. Both analyses were run using the Raup-Crick coefficient and a paired grouping algorithm in PAST (Hammer *et al.* 2001; Hammer & Harper 2006). Thanks to the large amount of data involved in these cluster analyses, it is possible to comment not only on the position of the Farewell Terrane, but also on the relationships of the other geographical entities.

The Sandbian–early Katian analysis (Fig. 5) shows a latitudinal gradient relative to the clustering of the different geographical entities. Of primary interest to this study, a strong cluster consisting of the Nixon Fork Subterrane and the Eastern Klamath Terrane is closely associated with Altai and Siberia, which again is tied to a cluster consisting of the shallow and deep-water faunas of cratonic Laurentia, the Midland Valley Terrane (Girvan, SW Scotland together with Kilbucho, Southern Uplands of Scotland and Pomeroy, Northern Ireland) together with a tight cluster of Avalonia and the deep-water faunas of Baltica. These clusters again are linked to the shallow-water faunas of South China and a cluster with the deep-water faunas of North- and South China and Sibumasu. This group of clusters show a weak affinity with the northern Precordillera, Kolyma, and the Cordillera Oriental. Thus, as Siberia has not been split up into a deep and shallow water component, due to lack of sufficient data, this analysis does not show how similar Siberian deep-water faunas are to the Nixon Fork Subterrane. However, from this analysis, it is clear that both the Eastern Klamath Terrane, as well as the Nixon Fork Subterrane, show stronger affinities with Siberia than Laurentia.

Overall this may be viewed as an equatorial cluster, consisting of a tropical platforms and margins together with tropical terranes; this is further associated with a tropical Gondwanan cluster. The shallow-water faunas of South China cluster here with deep-water faunas. This is due to the generally deeper facies of this faunally unusual palaeocontinent (Rasmussen & Harper, 2011a), which, earlier in the Ordovician probably initiated at least part of the Great Ordovician Biodiversification Event (Zhan & Harper 2006; Rasmussen *et al.* 2007) and in the Sandbian,

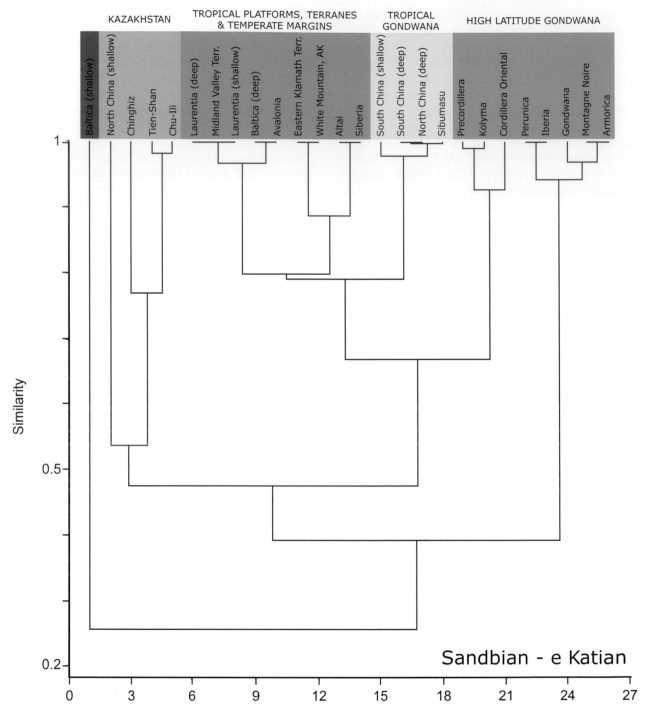

Fig. 5. Cluster analysis for the Sandbian–early Katian (Caradoc) interval based on 342 genera distributed across 26 locations.

witnessed the earliest appearance of the *Foliomena* fauna (Rong *et al.* 1999).

These two tropical clusters are related to a group of terranes located at higher latitudes, a cluster consisting of the Northern Precordillera and Cordillera Oriental. The position of Kolyma is not considered useful as too few species are known from this time slice, thus explaining its unexpected position in this diagram.

The 'Tropical clusters' and 'High latitude terranes' are together distantly related to the tight cluster of the Kazakh terranes which is also distantly related to the shallow-water faunas of North China. These are subsequently linked to a 'Gondwana' cluster of Perunica, Iberia, Gondwana, Montagne Noire and Armorica.

Finally, the shallow-water carbonates of Baltica appear surprisingly endemic with no apparent

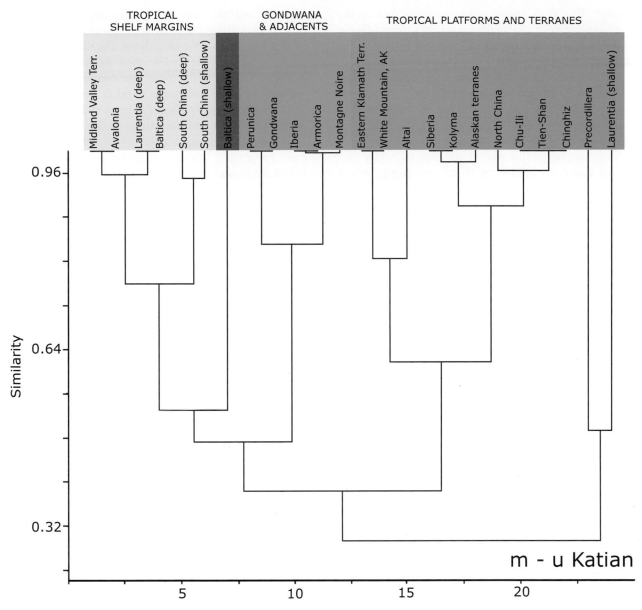

Fig. 6. Cluster analysis for the mid–early Katian (early–middle Ashgill) interval based on 289 genera distributed on 24 locations.

relationships to any of the other locations. The specific faunal links described in this study, i.e. *Oanduporella* and *Septomena*, probably belong to a deeper-water facies despite occurring in Estonia and Lithuania (Hints 1975; Paškevičius 1994; Rõõmusoks 2004).

This palaeogeographic setting had changed dramatically by the mid–late Katian, with brachiopod faunas becoming more and more cosmopolitan (Fig. 5). The Nixon Fork Subterrane and the Eastern Klamath Terrane now appear isolated and more weakly linked to the rest of the now 'Siberia and peri-Siberia' cluster. This cluster is essentially Siberia, Kolyma, North China (now undivided in facies due to lack of data) and the Kazakh terranes. Thus, within this mid–late Katian scenario, it seems that the Eastern Klamath

Terrane and the Nixon Fork Subterrane may have moved away from Siberia, compared to the Sandbian–early Katian interval. However, here it should be noted that the data from the Nixon Fork Subterrane is based on the less diverse localities 79WG19 and 79WG126. As already mentioned, the lack of sampled cosmopolitan taxa from Siberia, could explain why the Nixon Fork Subterrane does not cluster as tightly with this major palaeocontinent. Even so, the large Laurentian matrix does not seem to have particularly closer faunal affinities compared to the much smaller Siberian matrix. Instead the deep-water faunas of Laurentia cluster tightly with the deep-water faunas of Baltica and these have close affinities with the Midland Valley Terrane and Avalonia constituting a 'Tropical margins cluster'. This again shows moderate links to a cluster consisting of both the shallow and deep-water faunas of South China that arguably also could be part of the 'Tropical margins' cluster.

The shallow-water Baltic faunas are now no longer as isolated as earlier in the Ordovician, as these are now positioned between the 'Tropical margins' and Gondwana. The Gondwana cluster is essentially the same as that for the Sandbian–early Katian.

Surprisingly, in this matrix the shallow-water faunas of Laurentia appear to be an out-group, whose closest affinities are with the Northern Precordillera. The Precordillera drifted away from low latitudes and occupies a more isolated position in this analysis, probably due to its peri-Gondwanan location during the Katian, which was also demonstrated by Fortey & Cocks (2003) and Waisfeld, *et al.* (2003), both of which further argued for a close relationship between Laurentia and the Precordillera. Thus the result presented above is not surprising. However, it is surprising to see that the endemic character of the shallow-water Laurentian faunas show such a marked provincialism in the later parts of the Katian. This has not previously been demonstrated statistically.

In summary, the Nixon Fork Subterrane, and thus the Farewell Terrane, is closely associated with the Eastern Klamath Terrane and, in turn, Siberia with its associated terranes and to a lesser extent with cratonic Laurentia than previously believed. In the Katian matrix presented in Fig. 6, for instance, the Nixon Fork Subterranes faunal affinities with both the shallow and the deep-water Laurentian faunas are in fact the most unresolved in the entire data matrix. This can not be explained by insufficient Siberian data, nor do smaller samples from the Katian localities 79WG19 and 79WG126 explain it, since the largest data sets, with more than 100 species, are from the Laurentian data points. However, apart from the close Siberian affinity of the Farewell Terrane and the pronounced provincialism of the shallow-water Laurentian faunas, the broad trends developed in this analysis of Late Ordovician palaeogeography are consistent with the majority of previous studies (Torsvik *et al.* 1992; Fortey & Cocks 1998, 2003; Cocks & Torsvik 2002, 2007; Waisfeld *et al.* 2003; Torsvik & Cocks 2009).

Systematic palaeontology

Remarks. – The following chapter is a systematic overview of the taxa found within the material. Some taxa are only represented by a few shells or fragmented material and are only briefly described. These taxa have been included here to illustrate the full diversity of the fauna. The biogeographic distribution of the genera and subgenera described below, have been extracted from the database that forms the basis for the matrices used in Figures 5 and 6. Below, however, the Hirnantian interval has also been included. Ranges of the taxa are listed according to the global stages, unless if more detailed data are available. In those cases ranges are listed in brackets according to the British stage names.

Phylum Brachiopoda Duméril, 1806

Subphylum Linguliformea Williams, Carlson & Brunton, 1996

Class Lingulata Gorjansky & Popov, 1985

Order Lingulida Waagen, 1885

Superfamily Discinoidea Gray, 1840

Family Discinidae Gray, 1840

Genus *Acrosaccus* Willard, 1928

Type species. – *Acrosaccus shuleri* Willard, 1928; from the Rich Valley Formation (Sandbian) of Virginia, USA.

Biogeographic distribution. – Sandbian of Virginia (Willard 1928; Cooper 1956); upper Sandbian of the Oslo Region (Hansen 2008) and lower Katian of Kazakhstan (Popov & Cocks 2006).

Acrosaccus cf. *shuleri* Willard, 1928

Plate 1, figure 1

   1928 *Acrosaccus shuleri* sp. nov. Willard, p. 259, pl. 3, figs 1, 2.

Material. – Four ventral valves from locality A-1230, blocks 4–6.

Description. – Shell medium-sized, ventri-biconvex in profile and subcircular in outline. Ornamentation consists of concentric rugellae (concentric fila). Ventral valve conical, pedicle track narrow, closed by listrium. Foramen at posterior end of listrium and continued as internal tube. Interior unknown.

Remarks. – This species is referred to *A. shuleri* based on its outline. Apart from the type locality in Virginia, the species *A. willardi* is known elsewhere in North America from the upper Darriwilian of Alabama. The ventral valves of these two species are very similar, except for a more conical outline of *A. shuleri*. Other species within the genus are known from contemporaneous beds in Kazakhstan (Nikitin & Popov 1996; Popov & Cocks 2006) and from the Wenlock of the British Isles and Perunica (Mergl 2006).

Subphylum Rhynchonelliformea Williams *et al.*, 1996

Class Strophomenata Williams *et al.*, 1996

Order Strophomenida Öpik, 1934

Superfamily Strophomenoidea King, 1846

Family Strophomenidae King, 1846

Subfamily Strophomeninae King, 1846

Genus *Tetraphalerella* Wang, 1949

Type species. – *Tetraphalerella cooperi* Wang, 1949; from the upper part of the Elgin Member, Maquoketa Formation (middle Katian), Iowa, USA.

Biogeographic distribution. – Upper Darriwilian – lower Sandbian of the Nixon Fork Subterrane, west-central Alaska (Measures *et al.* 1992 and this study); lower Katian of the Chu-Ili Terrane (Nikitin *et al.* 2003); middle–upper Katian of Laurentia (Wang 1949; Howe 1988; Jin *et al.* 1997; Jin and Zhan 2001).

Tetraphalerella planobesa (Cooper, 1956), new combination

Plate 1, figures 2–8

> 1956 *Strophomena planobesa* Cooper, new species, p. 943, pl. 261, figs A1–13, B14–17.
> 1992 *Strophomena* sp. Measures, Rohr & Blodgett, p. 200, fig. 8G, H.

Material. – Two ventral valves from block 3 and 10 ventral and 11 dorsal valves from blocks 4–6, locality A-1230.

Description. – Shell gently biconvex, possibly longer than wide. Ventral valve more convex near the anterior margin; interarea low, wide and apsacline. Dorsal valve sulcate, apparently lacking interarea. Ornamentation unevenly multicostellate, with growth lines developed at every third of valve length.

Ventral interior with widely divergent dental plates. Median septum developed in a single specimen. Dorsal interior with bifid cardinal process extending posteriorly beyond hinge line. Socket ridges fused directly onto lateral bases of the cardinal process lobes. Concavity very pronounced anterior to cardinal process. Brachiophores low, supported by valve floor, widely divergent and slightly recurved towards the hinge line. Dorsal median septum lacking. Pair of low side septa present. Notothyrial platform absent.

Remarks. – Although no complete specimens are available for study, the current material is assigned to *T. planobesa* based on the relatively evenly costellate exterior, the only faintly developed dorsal median septum, the low, strongly apsacline ventral interarea and the low, wide pseudodeltidium – all characters that together separate this species from most other species of *Strophomena* and *Tetraphalerella*.

In addition, the species is transferred from *Strophomena* to *Tetraphalerella* based on the strongly undercut cardinal process with widely divergent socket ridges that are recurved towards the hinge line. True *Strophomena* has more anterolaterally directed socket ridges that give the dorsal cardinalia a triangular appearance (Cocks & Rong 2000). Whether or not to retain *Tetraphalerella* currently suppressed as a subgenus under *Strophomena*, as suggested in the Revised Treatise by Cocks & Rong (2000), or, to place it within its own family, as suggested by Dewing (1999), is difficult to assess based on the current material. Wang (1949) originally proposed to separate *Tetraphalerella* from *Strophomena* based on, among other characters, thinner socket ridges, more prominent and well-arranged pseudopunctae. The current material seems to possess a dorsal cardinalia that conform to Wang's definition, but as shown in the SEM image (Pl. 1, Fig. 8), not the well-arranged pseudopunctae. Dewing (1999, 2004) suggested, based on a study on the upper Katian *T. hecuba* and *S. planumbona*, that species with radially arranged, uncored pseudopunctae should be assigned to *Tetraphalerella*, whereas species with more irregularly arranged pseudopunctae with a smooth taleolate core, should be placed in *Strophomena*. The White Mountain species in the current study appears to have irregularly arranged pseudopunctae, as in *Strophomena*, but, as in *Tetraphalerella*, they appear to be lacking a taleolate core. Thus, a third option is that this may be an ancestral genus to the two. However, the current material is too fragmented to form the basis for a new genus.

Tetraphalerella was long thought to be endemic to Laurentia occurring in middle-upper Katian faunas, such as the Red River fauna. However, Nikitin *et al.*

(2003) reported an older species from the lower Katian part of the Tauken Formation in Kazakhstan. Thus, the current study may demonstrate that the oldest occurrence of the genus dates back to the early Sandbian.

Subfamily Furcitellinae Williams, 1965

Genus *Transridgeia* n. gen.

Type species. – *Transridgeia costata* n. sp.; from the upper Darriwilian – lower Sandbian of the Nixon Fork Subterrane, west-central Alaska.

Derivation of name. – Due to the very prominent set of transmuscle ridges in the dorsal valve interior.

Biogeographic distribution. – Upper Darriwilian-lower Sandbian of the Nixon Fork Subterrane, west-central Alaska.

Diagnosis. – Shell large, concavo-convex, wider than long, with subcircular outline. Interareas low, wide. Delthyrium high, moderately wide, covered completely by dental plates. Notothyrium wide, partly covered by apical chilidium. Ornamentation consists of 10–11 simple costae. Cardinal process bifid. Dorsal median septum surrounded by a complex arrangement of side septa and transverse ridges on a bema-like structure.

Description. – Shell large, markedly concavo-convex profile and transversely subcircular in outline. Maximum width at hinge line. Anterior commissure rectimarginate. Ventral valve unevenly convex, convexity strongest near anterior margin. Interarea low, wide and apsacline. Delthyrium high and moderately wide, covered completely by dental plates. Dorsal valve strongly concave, maximum convexity near anterior margin. Dorsal interarea low, flat and cataclinic. Notothyrium wide, partly covered by apical chilidium. Ornamentation consists of 10–11 simple costae.

Ventral interior with large teeth that are supported by widely divergent dental plates which extend posteriorly to cover delthyrium and converge onto valve floor to form a callus. Relatively deeply impressed, elongate, almost parallel diductor muscle scars are separated by a thin ventral median septum. Dorsal interior with widely divergent brachiophores almost parallel to hinge line. Cardinal process bifid; lobes are ventrally directed. Thin, low, dorsal median septum elevated from alveolus, continuing to about mid-valve length. Posteriorly, two pairs of transmuscle ridges are flanked medianly by a pair of side septa. The side

septa are flanked by perpendicular ridges that become wider anteriorly to form a triangular platform on both sides of the median septum in a bema-like structure. These triangular platforms are more elevated laterally away from the median septum.

Remarks. – This is the first record of this genus. However, the subfamily Furcitellinae has a near cosmopolitan distribution. The new genus is placed within the Furcitellinae due to its resemblance to *Dactylogonia* and other furcitellinids that have a dorsal median septum surrounded by a complex arrangement of side septa and transverse ridges. With respect to the dorsal median septum, the combination of an alveolus followed by a low, but sharply defined septum that terminates just anteriorly of the pronounced side septa is not seen in any other furcitellinids. Neither is the shape and outline of the side septa that are somewhat similar to those of *Glyptomena parvula* Cooper from the Effna-Rich Valley Formation in Virginia. That species differs in having a greater dorsal concavity in profile and more accentuated and wider separated pronounced costae and less well developed costellae. Interiorly, in the dorsal valve, the transmuscle ridges are longer in *G. parvula*. Further, within this subfamily, the ornamentation of simple costae only vaguely resembles that seen in *Panderites*, a genus whose outline and dorsal cardinalia is clearly separated from the new genus.

Transridgeia costata n. gen. *et* n. sp.

Plate 1, figures 9–13

Derivation of name. – Named for the prominent, wide costae developed on both valves.

Holotype. – MGUH 29451 (Pl. 1, fig. 12).

Paratypes. – MGUH 29448 (Pl. 1, fig. 9), MGUH 29449 (Pl. 1, figs 10, 13) and MGUH 29450 (Pl. 1, fig. 11).

Material. – Ten ventral and seven dorsal valves. All from locality A-1230, blocks 4–6.

Description. – As for the genus.

Remarks. – The three very pronounced pairs of side septa are quite distinctive, as is the shape of the bema-like structure.

Family Rafinesquinidae Schuchert, 1893
Subfamily Leptaeninae Hall Clarke, 1894

Genus *Leptaena* Dalman, 1828

Subgenus *Leptaena* (*Leptaena*) Dalman, 1828

Type species. – *Leptaena rugosa* Dalman 1828; from the Hirnantian *Dalmanitina* Beds of Västergötland, Sweden.

Biogeographic distribution. – The genus *Leptaena* is in need of revision (Cocks 2005). In the upper Ordovician numerous species have been described and erected, not least under open nomenclature. The oldest records of this genus are from the Darriwilian. Relevant to the biogeographic matrix presented in the current study, *Leptaena* is known from the lower Sandbian of Giuzhou, South China (Xu *et al.* 1974); lower–upper Sandbian of Ireland (Parkes 1992) and Wales (Williams 1963); upper Sandbian – lower Katian of the Oslo Region (Hansen 2008); lower Katian of Pomeroy, northern Ireland (Candela 2003), Kilbucho, Scotland (Candela & Harper 2010) and North China (Rong *et al.* 1999); the Katian of the Czech Republic (Havlíček 1982), France (Hammann *et al.* 1982; Mélou 1990), Kazakhstan (Rong *et al.* 1994) and Spain (Villas 1985); Streffordian–Cautleyan of Estonia (Rõõmusoks 1989) and Eastern USA (Patzkowsky & Holland 1997); Streffordian and Rawtheyan of Wales (Cocks & Rong 1988); upper Pusgillian – middle Rawtheyan of Scotland (Harper 2000, 2006); upper Cautleyan–Rawtheyan of Ireland, Québec and Sweden (Mitchell 1977; Sheehan & Lespérance 1979; Cocks 2005); Rawtheyan of Belgium (Sheehan 1987), and upper Rawtheyan of South China (Rong 1979, 1984a; b). From the Hirnantian it has been reported from Anticosti Island, Canada (Dewing 1999), the Czech Republic and Kazakhstan (Cocks 1988), Estonia (Rõõmusoks 1989), New Zealand (Cocks & Cooper 2004), Oklahoma, the Oslo Region, northeastern Russia (Kolyma), South China (Rong *et al.* 2002), Sweden (Bergström 1968; Dahlqvist *et al.* 2007) and Wales (Cocks 1988).

***Leptaena* (*Leptaena*) sp.**

Plate 1, figures 14–19

Material. – One pair of conjoined valves, four ventral and 19 dorsal valves. All from locality A-1230, blocks 4–6.

Description. – Medium to large sized shells, weakly planoconvex (geniculate), transverse, and strongly semi-oval. Maximum width at hinge line. About 50% as long as wide. Anterior commissure sharply geniculate dorsally. Ventral valve flat to weakly convex, strongly geniculate. Interarea moderately high, cataline. Delthyrium open, wide. Dorsal valve flat to weakly convex, strongly geniculate. Interarea low, flat, cataline to weakly anacline. Notothyrium open, wide. Ornamentation consists of concentric rugae and weakly impressed, closely spaced parvicostellae.

Ventral interior generally not preserved, however one fragment shows a small ventral median septum that extends anteriorly almost to the geniculate margin and at least one pair of muscle bounding ridges. Dorsal interior with flat, widely spaced brachiophores supporting probably a bifid cardinal process. Dorsal median septum weak continuing forward to about one-third of valve length as extension of ridge between brachiophores. A pair of short, side septa (muscle bounding ridges) are developed anterior of the hinge line, but do not fuse with the median septum. Muscle depressions are lobate, more deeply impressed posteriorly.

Remarks. – Shape and outline suggest this may be a new species. However, more material is needed to confirm whether this is a consistent feature.

Subgenus *Leptaena* (*Septomena*) Rõõmusoks, 1989

Type species. – *Leptaena juvenilis* Öpik, 1934 from the Kukruse Stage of Northern Estonia.

Biogeographic distribution. – Hitherto only known from the lower–upper Sandbian of northern Estonia (Rõõmusoks 2004).

***Leptaena* (*Septomena*) *alaskensis* n. sp.**

Plate 1, figures 20–23

Derivation of name. – Named after the sub-genus first occurrence in Alaska.

Holotype. – MGUH 29456 (Pl. 1, figs 21–22).

Paratypes. – MGUH 29455 (Pl. 1, fig. 20) and MGUH 29457 (Pl. 1, fig. 23).

Material. – Three ventral valves and 10 dorsal valves. All from blocks 4–6, locality A-1230.

Diagnosis. – Shell medium to large, plano-convex, subquadrate to slightly bilobate in outline; shell sharply geniculate dorsally. Maximum width at mid-valve. Anterior commissure uniplicate, sharply geniculate dorsally. Ventral interarea wide and high. Ornament of closely spaced parvicostellate intersected by larger costellae and concentric rugae. Dorsal interior with brachiophores flat and wide. Cardinal process bifid and undercut. Median septum weak with two pairs of transmuscle ridges. Inner pair subparallel, outer pair divergent, shorter, but more pronounced.

Description. – Shell medium to large, weakly plano-convex, subquadrate to slightly bilobate in outline; shell sharply geniculate dorsally. Maximum width at mid-valve. Anterior commissure uniplicate. Ventral valve flat to weakly convex, strongly geniculate. Inter-area moderately high, very wide, apsacline. Delthyri-um not preserved. Dorsal valve flat, strongly geniculate. Interarea not preserved and chilidium not present. Ornamentation of closely spaced parvicostel-lae intersected by larger costellae and concentric rugae.

Ventral interior not preserved, except for frag-ment showing interarea. Dorsal interior with flat, widely spaced brachiophores, sometimes parallel to hinge-line, supporting an undercut, bifid cardinal process. Dorsal median septum weak, with two pairs of transmuscle ridges. Inner pair subparallel to median septum, outer pair divergent, shorter, but more pronounced. Muscle depressions weakly impressed.

Remarks. – This species differs from *Leptaena* sp. in lacking a ventrally directed geniculation. The material is assigned to *L.* (*Septomena*) because of the presence of two distinctive pairs of transmuscle ridges. This is the first occurrence of the genus outside Estonia and the Alaskan material may be slightly older than that from Estonia. If so, this could indicate, as is the case with *Oanduporella*, that *Septomena* originated outside Baltica contrary to what was previously believed (Rõõmusoks 2004), or, conversely that Baltica may have been positioned relatively close to the Farewell Terrane during the early Sandbian and thereby, possi-bly, Siberia.

Family Christianiidae Williams, 1953

Genus *Christiania* Hall & Clarke, 1892

Type species. – *Leptaena subquadrata* Hall, 1883; from the lower Katian Tennessee, USA.

Biogeographic distribution. – This distinctive genus has been reported from the Sandbian of Kazakhstan (Popov *et al.* 2002); lower Sandbian of Altai (Senni-kov *et al.* 2008), Jones Ridge, east-central Alaska (Ross & Dutro 1966), eastern USA (Cooper 1956), Girvan (Williams 1962), Norway (Hansen 2008) and South China (Rong *et al.* 1999); upper Sandbian of Ireland (Mitchell 1977), New South Wales and Scotland (Per-cival 1991, 2009; Clarkson *et al.* 1992); upper

Sandbian – lower Katian of Perunica (Havlíček 1967), North China (Rong *et al.* 1999) and Thailand (Cocks & Fortey 1990), and in the lower Katian it is reported from Avalonia (Cocks 2010), Kazakhstan (Klenina *et al.* 1984; Popov *et al.* 2000; Popov & Cocks 2006), Pomeroy, northern Ireland (Candela 2003), Kilbucho, Scotland (Candela & Harper 2010), South China (Rong *et al.* 1999) and Taimyr (Cocks & Modzalevs-kaya 1997); Pusgillian–Rawtheyan of Girvan (Harper 2000, 2006) and South China (Rong *et al.* 1999); Raw-theyan of Belgium (Sheehan 1987); middle–upper Katian of Scania, Västergötland and Siljan, Sweden (Sheehan 1973, 1979) and Pomeroy, Ireland (Mitchell 1977); Rawtheyan of the Oslo Region (Cocks 1982), Poland (Cocks & Rong 1988), North Wales (Price 1981), Sardinia (Villas *et al.* 2002) and Maine, USA (Neuman 1994). Furthermore, the genus is reported from Ashgill rocks of Percé, Québec (Schuchert & Cooper 1930).

Christiania aseptata n. sp.

Plate 1, figure 24; Plate 2, figures 1–5

Derivation of name. – Refers to the weakly developed or entirely absent dorsal median septum.

Holotype. – MGUH 29461 (Pl. 2, fig. 5).

Paratypes. – MGUH 29458 (Pl. 1, fig. 24), MGUH 29459 (Pl. 2, figs 1, 2) and MGUH 29460 (Pl. 2, figs 3, 4).

Material. – 98 ventral and 61 dorsal valves from local-ity A-1230, blocks 4–6. In addition one dorsal frag-ment from locality 79WG19 and one dorsal fragment from locality 79WG126. These two Katian specimens may not be conspecific with the new Sandbian species.

Diagnosis. – Shell medium sized, concavo-convex in profile, elongate to subquadrate in outline. Delthyri-um open, partly covered by pseudodeltidium. Noto-thyrium open, wide. Chilidium absent. Valves are smooth but lamellose. Short ventral median septum. Dorsal median septum weak to absent.

Description. – Shell medium-sized with concavo-con-vex profile, elongate to subquadrate in outline. Maxi-mum width at hinge line. Anterior commissure rectimarginate to weakly unisulcate. Ventral valve strongly convex, especially near umbo. Interarea low, wide and strongly apsacline to orthocline. Delthyrium open, wide, partly covered by pseudodeltidium. Some specimens have a very small apical pseudodeltidium. Dorsal valve weakly concave. Interarea low, flat, weakly hypercline. Notothyrium open, wide. Chilidium absent. Valves are smooth but lamellose with minute growth lines variably developed.

Ventral interior with simple teeth supported by small dental plates. A short, small ventral median septum originates from delthyrium and diverges anteriorly to form a wide, low elevation that continues medially along entire valve length. Several pairs of side septa, or muscle bounding ridges are variably developed laterally. Muscle scars long, lobate, matching the position of the transmuscle ridges of the dorsal valve. Dorsal interior with flat, widely divergent brachiophores supporting a bifid cardinal process. Dorsal median septum very low, weak, if present. Possibly entirely absent. Deeply impressed bilobed muscle fields consist of two pairs of distinctive transmuscle ridges that each are fused anteriorly. In addition, a small septalium is situated within the muscle field about one-fifth of its entire length from the posterior margin.

Remarks. – Though the convexity of the ventral valve in strophomenoids is known to show a substantial variation at the species level in some populations (Rong & Cocks 1994), the Alaskan species of *Christiania* is separated from *C. subquadrata* Hall by the deeper ventral valve and the smaller ventral interarea, a delthyrium not entirely covered by pseudodeltidium, and the absent to very weak dorsal median septum. A great number of species have been erected within this genus, thus making it difficult to get a morphological overview of the differences that separates these species. However, following Popov & Cocks (2006), a key feature is the relative length of the dorsal median septum. As they erected a new species, *C. proclivis*, on the basis of a dorsal median septum that is less than half of the valve length, it seems reasonable to separate the Alaskan species, which almost lacks a dorsal median septum, from species with a septum, as well as from *C. proclivis*. Furthermore, the Alaskan species appears more transverse than *C. proclivis*. In addition, the ventral median septum is larger and the prominent side septa are more parallel unlike the posteriorly inclined side septa in *C. proclivis*. Some species, such as *C. egregia* Popov (1985) and *C. sulcata* Williams (1962), are noted for their ventral median sulcus. The Alaskan species shows some variation in external morphology as some specimens do have a ventral median depression, but others do not. However, internally the Kazakh *C. egregia* resembles *C. aseptata* in having more or less parallel side septa in the dorsal valve. But, as with the ventral depression, this feature seems to vary within this relatively large material of *C. aseptata*, though never becoming as divergent anteriorly as that seen in *C. proclivis*. Instead, the new species is clearly separated from *C. egregia* by having a much shorther ventral valve.

Family Unknown

Strophomenoidea gen. *et* sp. indet.

Plate 2, figures 6–7

Material. – One ventral and five dorsal valves. All from locality A-1230, blocks 4–6.

Description. – Shell dorsi-biconvex, subcircular in outline. Ventral valve gently and evenly convex; interarea not preserved. Dorsal valve more convex near the posterior margin; dorsal interarea not preserved. Ornamentation is unevenly parvicostellate, with 2–3 smaller costellae separated by one or more accentuated round costellae.

Ventral interior not well preserved, except for widely divergent dental plates. Dorsal interior with large, bifid cardinal process, which bases are joined together in a keel anterior of the hinge line. Cardinal process extends posteriorly beyond hinge line, where it becomes wider; supported by socket ridges, crenulated anteriorly. Pair of side septa extends the full length of the valve to the anterior margin.

Remarks. – The keeled cardinal process is reminiscent of the Triplesiidae. However, the appearance of the cardinal process may be obscured due to the poorly preserved shells. Further, the ornamentation is not typical of the Triplesiidae. Therefore, these specimens are not referred to a family.

Strophomenoidea gen. *et* sp. indet. 2

Plate 2, figures 8–10

Material. – 19 ventral and one dorsal. All from locality A-1230, blocks 4–6.

Description. – Shell large, plano-convex, geniculate; outline transversely subrectangular. Ventral valve weakly and unevenly convex near umbonal region; interarea low, wide and cataline. Dorsal interarea unknown. Ornamentation parvicostellate with thin, widely spaced costae. One accentuated costae for every two costellae. Interiors not well preserved. Ventral valve with teeth that are small, low and wide, not supported by dental plates.

Remarks. – A thorough description of the species is not possible due to the poor preservation of the material. However, morphological characters do suggest a new genus. More material with cardinalia is needed to enable the possible erection of a new genus. The epifauna seen on Plate 2, figure 7 consists of *Ptychopleurella uniplicata* (same specimen is also illustratated in Figure 4). This may suggest that *Ptychopleurella* has not been transported far, and thus belongs to the deep water assemblages. Conversely, this large strophomenoid shell could well be derived from a more shallow-water setting. Furthermore, borings are seen in some valves.

Strophomenoidea gen. *et* sp. indet. 3

Plate 2, figures 11–12

Material. – One ventral fragment from block 3, locality A-1230.

Description. – Outline and profile incompletely preserved, but the valve appears to have a long, high ventral interarea that is either procline or orthocline. Umbonal region is probably positioned posterior of hinge-line. Ornament consists of widely spaced costae (about 1/mm). Teeth large. Interiors badly preserved.

Remarks. – This sole ventral fragment is assigned to the strophomenoids on the basis of its characteristic ornament.

Strophomenoidea gen. *et* sp. indet. 4

Plate 2, figures 13–18

Material. – Nine ventral and five ?dorsal fragments from blocks 4–6, locality A-1230.

Description. – Shell large with strongly convexo-concave profile; outline incomplete, but probably with obtuse cardinal angles. Ventral valve shallowly convex. Maximum convexity at umbo. Valve strongly resupinate towards the anterior margin. Dorsal valve flat almost up to anterior commissure, where it becomes strongly resupinate ventrally. Ornamentation of fine, closely spaced costellate, separated by a single more accentuated costae every 1.5 mm (measured at the anterior commissure). Strong, closely spaced rugae, perpendicular to the radial ornament. Ventral interarea high, wide and cataclime. Dorsal interarea, wide, probably anacline.

Teeth large, wide and not supported by dental plates. Ventral muscle field broadens anteriorly, confined laterally by a muscle bounding ridge, more elevated posteriorly. Small ridges developed medially. Additional ridge developed laterally. This may, however, be a preservational artefact. Cardinal process not preserved. Brachiophores broad and crenulated.

Medianly, dorsal median septum does not seem to be a continuation of the cardinal process. Median septum originating in front of an alveolus, as a low elevated median ridge and flanked by two side septa confining a muscle field that becomes progressively narrower anteriorly.

Remarks. – Unfortunately, the specimens assigned to this species are heavily fragmented and the dorsal interior is based on material that is only questionably assigned to this genus. The unusual combination of ornamentation and profile does indicate that a new genus could be erected. The shape suggests that the taxon may be related to *Leptaena* based on its convexo-concave profile combined with the characteristic rugate ornamentation. But in fact, the ornamentation is more similar to that of the Baltic endemic *Gunnarella* (Spjeldnæs 1957). Both are resupinate, but different from specimens described here. The ventral muscle field resembles that of typical strophonellids (see Rong & Cocks 1994, text-fig. 9E)), whereas, according to Rong & Cocks (1994), the combined structure of a dorsal transmuscle ridge and side septa is more typical of *Quondongia* Percival (1991).

Superfamily Plectambonitoidea Jones, 1928

Family Sowerbyellidae Öpik, 1930

Subfamily Sowerbyellinae Öpik, 1930

Genus *Sowerbyella* Jones, 1928

Subgenus *Sowerbyella* (*Sowerbyella*) Jones, 1928

Type species. – *Leptaena sericea* Sowerby, 1839; from the lower Sandbian of Shropshire, England.

Biogeographic distribution. – Reported from the lower Sandbian of England (Sowerby 1839), Myanmar (Cocks & Rong 1989) and Wales (Williams 1963); lower Sandbian – lower Katian of New South Wales, Australia (Cocks & Rong 1989; Percival 1991); lower Sandbian – upper Katian of the Chu-Ili Range (Popov *et al.* 2000, 2002; Nikitin *et al.* 2003) and the Chinghiz Range (Klenina *et al.* 1984; Popov & Cocks 2006), both Kazakhstan, Estonia (Öpik 1930; Rõõmusoks 1959), Ireland including Pomeroy (Mitchell 1977; Parkes 1992; Candela 2003) and Eastern USA (Patzkowsky & Holland 1997); upper Sandbian of Norway (Spjeldnæs 1957); upper Sandbian – lower Katian of eastern Canada (Schuchert & Cooper 1930); upper Sandbian–Rawtheyan of Girvan (Cocks & Rong 1989;

Harper 2006); Katian of Kolyma (Cocks & Rong 1989) and Sweden (Angelin & Lindström 1880; Cocks 2005); lower Katian of Kilbucho, Scotland (Clarkson *et al.*1992; Candela & Harper 2010), Kyrgyzstan (Popov *et al.* 2000; Nikitin *et al.* 2003) and Mongolia (Rozman 1978); lower–middle Katian of North China (Cocks & Rong 1989; Rong *et al.* 1999); upper Cautleyan of Québec (Sheehan & Lespérance 1979); Rawtheyan of South China (Rong *et al.* 1999), Uzbekistan (Nikiforova 1978), Taimyr (Cocks & Modzalevskaya 1997) and Wales (Jones 1928) and finally the Hirnantian of the Oslo Region (Baarli 1995).

Sowerbyella (Sowerbyella) rectangularis n. sp.

Plate 2, figures 19–24

Derivation of name. – Referring to the new species rectangular outline, compared to other known species within the genus.

Holotype. – MGUH 29473 (Pl. 2, figs 19, 21).

Paratypes. – MGUH 29474 (Pl. 2, fig. 20), MGUH 29475 (Pl. 2, figs 22, 23) and MGUH 29476 (Pl. 2, fig. 24).

Material. – Two ventral valves from block 3 and 22 pairs of conjoined valves, 32 ventral and 42 dorsal valves from blocks 4–6, locality A-1230. One ventral valve from locality 79WG126.

Diagnosis. – Shell small, moderately to weakly concavo-convex, with transverse outline. Interareas low, covered partly by pseudodeltidium and chilidial plates, respectively. Dense, unequally parvicostellate ornament separated by 12 to 14 accentuated costae. Ventral muscle field bilobed, cordate. Short, thick ventral median septum. Cardinal process trifid, undercut. Dorsal median septum absent.

Description. – Shell moderately to weakly concavo-convex, transverse, rectangular to semi-oval in outline. About twice as wide as long, with acute cardinal extremities. Anterior commissure rectimarginate. Ventral valve evenly convex with a low, wide apsacline interarea covered apically by pseudodeltidium. Dorsal valve moderately to gently concave, with a flat, hypercline interarea and discrete chilidial plates. Radial ornament unequally parvicostellate with 7–10 parvicostellae per mm along anterior margin of mature specimens and with 12–14 accentuated ribs at umbo.

Ventral interior with small short teeth and divergent dental plates. Ventral muscle field bilobed, cordate, slightly transverse, about one-third as long as

valve length. Posteriorly, a small, short, but thick ventral median septum separates moderately impressed elongate diductor scars that enclose small, lanceolate adductor scars. Mantle canals system not seen. Dorsal interior with undercut, trifid cardinal process. Median septum absent, instead a pair of narrow-angled side septa are developed and extends for about 2/3 of valve length. Additional side septa or adductor muscle bounding ridges occur.

Remarks. – The specimens from locality A-1230 are flatter than other known species of *Sowerbyella*, notably the more convex type species, *S. sericea* (Sowerby 1839). In addition, the new species described here appears more rectangular in outline with almost right-angled cardinal angles as opposed to the subcircular outline with acute cardinal angles of the type species, as well as *S. thraivensis* (Reed 1917) and *S. sladensis* Jones 1928;. Zhan & Cocks (1998), also reported a species, *S. sinensis*, from the Upper Ordovician of China that is also less convex than the type species. However, the Alaskan species is differentiated from *S. thraivensis* as well as the type species, among others, in having a short, but thick, ventral median septum. *Sowerbyella nativa* Klenina *et al.* 1984 from the middle Ordovician of Kazakhstan is similar in outline to the Alaskan species, however, its profile is much more convex. The dorsal interior shows very pronounced side septa with no apparent bema or median septum. The central pair of side septa is the longest. Anteriorly it is positioned on an elevated thickening of the valve. The bema is well developed in the type species, but is known to be variably developed, or absent, within the genus. The lack of a median septum is also common within this genus, and this separates the Alaskan species from *S. rukavishnikovae* Popov *et al.* 2002 from the Upper Ordovician of Kazakhstan. In addition, *S. rectangularis* has more densely developed costellae.

The specimen from locality 79WG126 is too poorly preserved to compare with the rest of the material.

Subgenus *Sowerbyella (Rugosowerbyella)* Mitchell, 1977

Type species. – *Plectambonites subcorrugatella* Reed 1917 from the middle Katian (Cautleyan) Killey Bridge Formation of Pomeroy, Northern Ireland.

Biogeographic distribution. – *Rugosowerbyella* is reported from the upper Sandbian of Pomeroy (Candela 2003); upper Sandbian–Hirnantian of Kazakhstan (Klenina *et al.* 1984; Cocks 1988; Popov & Cocks 2006); early Katian of Estonia (Cocks 2005); Katian of

Taimyr (Cocks & Modzalevskaya 1997); Streffordian–Cautleyan of the Midland Valley Terrane (Girvan, Harper 2006); Pusgillian–Rawtheyan of Sweden (Cocks & Rong 1989; Cocks 2005; Zhan & Jin 2005); Cautleyan of Pomeroy, Ireland (Mitchell 1977); Rawtheyan of Belgium (Sheehan 1987), Poland (Mergl 1990) and South China (Zhan & Cocks 1998).

Sowerbyella (*Rugosowerbyella*) *praecursor* n. sp.

Plate 3, figures 1–5

Derivation of name. – The species name refers to this being the earliest known occurrence of the genus.

Holotype. – MGUH 29478 (Pl. 3, figs 2, 5).

Paratypes. – MGUH 29477 (Pl. 3, figs 1, 4) and MGUH 29479 (Pl. 3, fig. 3).

Material. – Four pairs of conjoined valves, five ventral valves and seven dorsal valves from blocks 4, 5 and 6. All from locality A-1230.

Diagnosis. – Shell shallowly concavo-convex in profile, and transversely semi-oval in outline. Small pseudodeltidium. Ornament unequally parvicostellate interrupted by weakly developed rugae. Ventral muscle field bilobed, cordate with very short median septum. Dorsal interior with a trifid, undercut cardinal process, no septum present. Two pairs of side septa developed. The proximal pair extends farthest anteriorly.

Description. – Weakly concavo-convex, transverse and semi-oval in outline. About twice as wide as long, with acute cardinal angles. Anterior commissure rectimarginate. Ventral valve evenly convex, with a low, wide apsacline interarea. Small pseudodeltidium may be present apically. Dorsal valve gently concave, with a low, wide, flat and cataclinetohypercline interarea. Ornament of concentric rugae, becoming more accentuated towards the umbonal zone. Radial ornament unequally parvicostellate with 3–5 parvicostellae per mm along anterior margin of mature specimens and with 6–10 accentuated costae at anterior commissure.

Ventral interior with small teeth and short, divergent dental plates. Ventral muscle field bilobed, cordate, slightly transverse, about one-fifth as long as valve length though in some specimens extends to about mid-valve. Very short ventral median septum. Adductor scars or mantle canals system not impressed. Dorsal interior with a trifid, undercut cardinal process. Adductor scars enclosed posteriorly by lateral myophore ridges and two pairs of side septa.

The pair most proximal to the median extends farthest anteriorly. No dorsal median septum observed.

Remarks. – This species differs from *Sowerbyella* (*S.*) *rectangularis* primarily in having distinctive concentric rugae. Internally the lack of a dorsal median septum and a more narrow-angled pair of side septa medially helps separate the two subgenera from each other and from *Eoplectodonta*. The current species is distinguished from other species within this genus in its more weakly developed rugae and fewer, more widely spaced pronounced costae. Further, it is separated from the type species in lacking a bema. One may speculate that the development of more pronounced rugae and a bema may be a phylogenetic adaptation. However, as a number of the morphological features are not typical of *Rugosowerbyella*, this species may be described as a new genus.

Sowerbyella (*Rugosowerbyella*) sp. 1

Plate 3, figures 6–7

Material. – Five ventral fragments and one indeterminable fragment from block 3, A-1230.

Description. – Valves broken; outline and profile hard to determine; ventral valve may be weakly convex. Umbo just posterior of hinge line. Ventral interarea high. Ornament is unequally parvicostellate with about 10–13 costellae between accentuated costa. Rugae are present on entire valve perpendicular to the radial ornamentation. Ventral interior with median septum and deeply impressed muscle scars. Thick, wide dental plates.

Remarks. – This species is possibly conspecific with *R. praecursor*. However it possesses a different, more deeply impressed ventral muscle scar, a longer ventral median septum and stronger dental plates. The ventral interarea is also larger and higher. The costae are more strongly impressed. These features, however, may be a result of the preservational state of this fragmented and very limited material.

Genus *Anisopleurella* Cooper, 1956

Type species. – *Anisopleurella tricostellata* Cooper 1956 from the upper Darriwilian Pratt Ferry Formation of Alabama, USA.

Biogeographic distribution. – This genus is known from the lower Sandbian – upper Katian of Girvan (Williams 1962; Harper 2006), South China and Wales (Cocks & Rong 1989); upper Sandbian – lower

Katian of Kazakhstan (Klenina *et al.* 1984), North China (Rong *et al.* 1999) and the Oslo Region (Hansen 2008); upper Sandbian to middle Katian of Ireland (Mitchell 1977; Parkes 1992; Candela 2003); Pusgillian–Cautleyan of South China (Zhan & Jin 2005); upper Cautleyan–Rawtheyan of Sweden (Cocks 2005); Rawtheyan of Belgium (Sheehan 1987), Maine, USA (Neuman 1994) and of the Czech Republic (Havlíček 1967); Rawtheyan–Hirnantian of Wales (Jones 1928; Cocks & Rong 1988) and additionally it is reported from the Hirnantian of Kazakhstan (Cocks 1988).

Anisopleurella tricostata n. sp.

Plate 3, figures 8–12

Derivation of name. – Emphasizes the new species apparently smooth exterior that is intersected by three distinct costae.

Holotype. – MGUH 29481 (Pl. 3, fig. 8).

Paratypes. – MGUH 29482 (Pl. 3, fig. 9), MGUH 29483 (Pl. 3, figs 10, 11) and MGUH 29484 (Pl. 3, fig. 12).

Material. – Five pairs of conjoined valves, 56 ventral and 10 dorsal valves. All from locality A-1230, blocks 4–6.

Diagnosis. – Shell moderately concavo-convex, with transverse outline, twice as wide as long. Cardinal angles acute. Apical pseudodeltidium possibly present. Chilidium present. Faintly developed unequally parvicostellate ornamentation with three prominent costae. Angle of 45 degrees between two consecutive costae. Ventral interior with small teeth, not supported by dental plates. Dorsal interior with undercut, probably bilobed, cardinal process. Socket ridges wide, curving back laterally towards hinge line. Anteriorly elevated dorsal median ridge and pair of prominent side septa. Side septa bisecting, but confined within a bilobed, subelliptical bema.

Description. – Shell moderately concavo-convex, with transverse to semi-oval outline, about twice as wide as long, and with acute cardinal angles. Anterior commissure rectimarginate. Ventral valve strongly convex towards anterior margin, with low, wide, apsacline interarea. Small apical pseudodeltidium possibly present, however difficult to determine with certainty due to the preservational condition of material. Dorsal valve moderately concave, with a low, wide flat hypercline interarea. Chilidium present. Ornamentation unequally parvicostellate, usually with three prominent costae;

angle of 45 degrees between two consecutive costae. Costellae are closely spaced, but very weakly developed.

Ventral interior with small teeth, not supported by dental plates. Ventral muscle field weakly impressed, but apparently bilobed. A very weak median septum or bulge, present in some specimens. Dorsal interior with undercut, bilobed cardinal process, extending posterior to hinge-line. Socket ridges very wide, curving back laterally towards hinge line. Dorsal median ridge and pair of prominent side septa more elevated anteriorly. Side septa bisecting, but confined within a bilobed, subelliptical bema. Bema extending forward to about three-quarters of valve length. A series of ridges are developed anteriorly between the two lobes of the bema.

Remarks. – The ornament, outline and profile all are diagnostic of *Anisopleurella*. Internally, the dorsal bema is also diagnostic of this genus. The new species is separated from other known species by its almost entirely smooth exterior apart for the three more prominent costae. The external surface of the shell consists of very closely spaced, minute costellae. They are so faintly developed that the surface, in most specimens, appears smooth if not observed under high magnification. The type species *A. tricostellata* Cooper 1956 was also named after its three prominent costellae. However, whereas the new species externally appears smooth apart from the three costae, the type species depicted by Cooper (1956), pls 193A, 195A) often has more than three prominent costellae that intersects well developed, closely spaced costellae. Further, the new species has a more convex ventral valve and a less transverse outline than seen in the type species. Cooper also erected *A. inaequistriata* Cooper 1956, which has several prominent costellae and has an even more transverse outline than the type species. Internally, in the dorsal valve, the new species is unique in having a series of ridges developed anteriorly between the two lobes of the bema and the anterior margin.

Subgenus *Eoplectodonta* (*Eoplectodonta*) Kozłowski, 1929

Type species. – *Leptaena duplicata* Sowerby, 1839; from the Lower Llandovery (Rhuddanian) Gouleugoed Formation of Llandovery, Wales.

Biogeographic distribution. – This subgenus has been reported from the lower Sandbian of eastern USA (Cocks & Rong 1989) and South China (Rong *et al.* 1999); Sandbian of Wales (Williams 1963); lower Sandbian – lower Katian of the Oslo Region (Hansen 2008); upper Sandbian of Scotland (Clarkson *et al.* 1992); upper Sandbian – lower Katian of North

China (Zhan & Jin 2005), Sardinia, Italy and Tien-Shan (Cocks & Rong 1989); upper Sandbian – upper Katian of Girvan (Harper 2000, 2006) and Pomeroy, Ireland (Mitchell 1977; Candela 2003); lower Katian of British Columbia (Jin & Norford 1996), Chu-Ili Terrane (Nikitin *et al.* 2006) and the Mongolian Altai (Cocks & Rong 1989); Katian of Taimyr, North Russia (Cocks & Modzalevskaya 1997), Kazakhstan (Klenina *et al.* 1984) and South China (Rong *et al.* 1999; Chen *et al.* 2000); upper Cautleyan–Rawtheyan of Québec (Sheehan & Lespérance 1979) and Sweden; Rawtheyan of Poland (Zhan & Jin 2005) and Wales (Cocks 2005); Rawtheyan–Hirnantian of England (Cocks & Rong 1989; Cocks 2005) and Norway (Cocks 1988, 2005; Cocks & Rong 1989). In addition, this subgenus is also reported from the Hirnantian of New Zealand (Cocks & Cooper 2004), Kazakhstan and northeast Russia (Oradovskaya 1983; Cocks 1988).

Eoplectodonta (*Eoplectodonta*) sp.

Plate 3, figures 13–15

Material. – Two fragments from locality A-1230 from block 3 and 12 dorsal valves from blocks 4–6. Two dorsal valves from locality 79WG126.

Description. – Shell strongly concavo-convex in profile, transverse to suboval in outline and with acute cardinal angles. Anterior commissure rectimarginate. Ventral valve not identified with certainty. Ornamentation unequally parvicostellate, usually with more prominent costa every 20°. Rugae are developed distally along the posterior margin.

Dorsal interior with erect, trifid, undercut cardinal process. Socket ridges 'triangular' as compared to *Rugosowerbyella* in which these are more 'bent' or curved. Two pairs of relatively prominent side septa originate just anterior of the cardinal process and extend forward for at least 80% of valve length. About mid-valve, a less prominent, delayed dorsal median septum starts, ending at about the same position as the side septa relative to the anterior margin. Denticles not preserved on hinge-line.

Remarks. – This species strongly resembles *Sowerbyella* and *Rugosowerbyella*. In this study the distinction is primarily based on the delayed dorsal median septum in *Eoplectodonta*. The material most likely also includes ventral valves that could be assigned to this genus. However, possibly due to the fragmented state of the material, no denticles are observed on any valves. Therefore, all relevant ventral valves in the material are referred to either *Sowerbyella* or

Rugosowerbyella. Further, the relevant ventral valves in the current material all possess a small, posteriorly positioned ventral median septum, a feature that is usually absent in *Eoplectodonta* (Cocks & Rong 1989).

Subfamily Ptychoglyptinae Cooper, 1956

Genus *Ptychoglyptus* Willard, 1928

Type species. – *P. virginiensis* Cooper, 1956; from the lower Sandbian Edinburg Formation of Virginia, USA.

Biogeographic distribution. – Reported from the Sandbian of Avalonia (Parkes 1992; Cocks 2010); lower Sandbian of Argentina, Girvan and Virginia, eastern USA (Cocks & Rong 1989), the Southern Shan States, Myanmar (Cocks & Zhan 1998); Sandbian–middle Katian of Kolyma (Oradovskaya 1983; Cocks & Rong 1989); upper Sandbian of Girvan and upper Sandbian – lower Katian of Norway (Cocks & Rong 1989; Hansen 2008) and Kazakhstan (Klenina *et al.* 1984); lower Katian of Sweden (Cocks 2005) and Pomeroy, Northern Ireland (Candela 2003); Pusgillian of Girvan, Scotland (Harper 2006); upper Cautleyan–Rawtheyan of Sweden (Cocks 2005); Rawtheyan of Belgium (Sheehan 1987); the Hirnantian of Quebec (Schuchert & Cooper 1930). Furthermore, Ross & Dutro (1966) reported *Ptychoglyptus* from upper Ordovician rocks of Jones Ridge, east central Alaska.

Ptychoglyptus pauciradiatus Reed, 1932

Plate 3, figures 16–19

1932 *Ptychoglyptus pauciradiatus* Reed, pp. 122–123, pl. XVIII, figs 1, 2.
1966 *Ptychoglyptus*? cf. *pauciradiatus* Reed; Ross & Dutro, p. 17, pl. 3, figs 18, 19.

Material. – Nine ventral fragments, six dorsal fragments and one indeterminate fragment. All from blocks 4, 5 and 6, locality A-1230.

Description. – Shell medium sized, equibiconvex, transversely semi-oval in outline. Maximum width at hinge-line. Acute cardinal angles. Anterior commissure rectimarginate. Ventral valve weakly convex, with a low, wide, strongly apsacline interarea. Small apical pseudodeltidium and apical foramen present in some specimens. Dorsal valve flat, but weakly concave posteromedianly. Interarea flat, low, wide and orthocline. Chilidium closed by chilidial plates. Ornamentation consists of distinctive, small, closely spaced rugae, interrupted by closely spaced weak costellae and up to seven more distinctive costae.

Ventral interior with small teeth supported by short recessive, widely divergent dental plates. Ventral muscle field weakly impressed, but probably bilobed. Median septum lacking, but elevated area bisects the muscle field. Dorsal interior with a cardinal process that may be undercut, however, this is uncertain due to the preservational state of the specimens. Socket ridges long and wide. Median septum with 2 or 3 narrow-angled side septa present.

Remarks. – Ross & Dutro (1966) assigned their specimens from the Jones Ridge area, east-central Alaska, to *P. pauciradiatus*; a species also recorded from the Upper Allochthon of the Caledonian nappes in the Trondheim region of Norway (Reed 1932). These have previously been interpreted as island arcs originating from within the Iapetus Ocean, somewhere close to Laurentia during the Ordovician (Harper *et al.* 1996). Ross & Dutro's (1966) identification of this species was based on similar concentric ornamentation as well as the existence of seven primary costae. As the species from the White Mountain area possess the same number of primary costae, it is also assigned, to *P. pauciradiatus* Reed. This species, however, is transversely semi-oval, whereas the Jones Ridge material (one specimen) is more rectangular, i.e. wider than long. Ross & Dutro (1966) further questioned whether it may be conspecific to some of Williams's (1962) specimens of *Glyptambonites* aff. *glyptus* from the Sandbian of Girvan. The White Mountain material differs in both radial and concentric ornament from *Glyptambonites* aff. *glyptus*. The current species could also be conspecific with *Ptychoglyptus alaensis* n. sp. (see below), in which case *P. pauciradiatus* could be juvenile specimens of *P. alaensis*. However, the lack of alae, more closely spaced rugae, flatter ventral valve and an apparently undercut cardinal process seem to differentiate this species from *Ptychoglyptus alaensis* n. sp.

Ptychoglyptus alaensis n. sp.

Plate 3, figures 20–24

Derivation of name. – The name emphasizes the strongly alate outline of the new species.

Holotype. – MGUH 29490 (Pl. 3, figs 20, 22).

Paratypes. – MGUH 29491 (Pl. 3, figs 21, 23) and MGUH 29492 (Pl. 3, fig. 24).

Material. – Nine ventral and seven dorsal valves, all from blocks 4–6, locality A-1230.

Diagnosis. – Shell large, equally biconvex in profile. Transverse to suboval outline. Cardinal extremities alate. Hinge line bends laterally in anterior direction. Ventral interarea high, cataline. Delthyrium partially covered. Dorsal interarea low, orthocline to hypercline. Notothyrium covered by chilidium. Ornament of numerous thin costellae interrupted by accentuated costae and broad, wavy rugae. Ventral interior with small teeth, not supported by dental plates. Dorsal interior with a trifid, undercut cardinal process. Socket ridges long and wide. Very short, posteriorly developed, median septum with one pair of narrow-angled side septa developed as an anterior extension of the median septum.

Description. – Shell large, equi-biconvex in profile, with dorsal valve slightly concave near the posterior margin. Outline transverse to suboval; cardinal angles alate. Hinge line bends laterally in an anterior direction. Ventral interarea wide, high and cataline. Delthyrium partially covered. Dorsal interarea low, wide and orthocline to weakly hypercline. Notothyrium covered by chilidium. Ornamentation of numerous thin costellae interrupted by one more accentuated costae for every 10°. Rugae perpendicular to radial ornament, interrupting it; rugae broad and wavy, impressed on the valve interiors.

Ventral interior with low, divergent dental plates and divergent muscle field. Teeth small compared to size of interarea. Dorsal interior with cardinalia resembling *P. pauciradiatus* but the cardinal process continues as a low, thick median ridge possibly with a pair of narrow spaced side septa. Laterally, next to side septa, deep depressions are seen. Socket ridges widely divergent; sockets deep and large.

Remarks. – This species is differentiated from *P. pauciradiatus* on the basis of its larger size, alate outline, a more flat profile, curved hinge line and broader more wavy rugae than in *P. pauciradiatus*. The rugae more resemble that of *Leptaena*. The rounded outline with the very distinctive, broader alae is not known from any other species within this genus.

Family Syndielasmatidae Cooper, 1956

Genus *Sowerbyites* Teichert, 1937

Type species. – *S. medioseptatus* Teichert, 1937; from the Upper Ordovician Limestone, Ignertoq, east coast of Melville Peninsula, Arctic Canada.

Biogeographic distribution. – This genus has been reported from the upper Ordovician of Arctic Canada (Teichert 1937); lower Sandbian of Eastern and Central USA (Cocks & Rong 1989); Sandbian–lower Katian of Kazakhstan (Kulkov & Severgina 1989) and Australia (Laurie 1991; Percival 1991); upper Sandbian and lower Katian of Pomeroy, Ireland (Mitchell 1977; Candela 2003) and Scotland including Kilbucho (Clarkson *et al.* 1992; Candela 2006a; Candela & Harper 2010); lower Katian of Mongolia (Cocks & Rong 1989) and Russian Altai (Rong *et al.* 2007).

Sowerbyites sp.

Plate 4, figures 1, 2

Material. – One dorsal valve from locality A-1230, blocks 4–6.

Description. – Shell gently concave, transverse, semi-elliptical in outline. Cardinal angles obscured. Anterior commissure rectimarginate, strongly lamellose. Ornament unequally parvicostellate with accentuated costae relatively closely spaced. Dorsal interior probably with thick, anteriorly directed socket ridges. Median septum thick, moderately elevated and extends almost to the anterior margin. One pair of side septa present.

Remarks. – According to Cocks & Rong (1989), the interior of the dorsal valve has only been illustrated in one specimen of the type species. Therefore, it is difficult to assess the variability of this genus. Moreover, following the same authors, the presence of side septa would exclude this specimen from the Leptellinidae and instead place it within the Syndielasmatidae. A dorsal median septum further separates this genus from *Syndielasma* Cooper 1956, the only other genus within the family.

In the current species the dorsal median septum is elevated anteriorly and the side septa is well developed – all these characters suggest assignment to *Sowerbyites*.

Family Bimuriidae Cooper, 1956

Genus Bimuria (Bimuria) Ulrich & Cooper, 1942

Type species. – *Bimuria superba* Ulrich & Cooper 1942; from the upper Darriwilian Middle Arline Formation, Tennessee, USA.

Biogeographic distribution. – This genus has previously been reported from the lower Sandbian of Chinghiz, Kazakhstan (Nikitin & Popov 1984), Alabama and Virginia, USA (Cooper 1956); Sandbian of Avalonia (Williams 1962) and Jones Ridge, east-central Alaska (Ross & Dutro 1966); lower Sandbian to middle Katian of Girvan (Williams 1962; Harper 1989; Candela & Harper 2010) and Pomeroy, Ireland (Mitchell 1977; Candela 2003); upper Sandbian of Ireland (Parkes 1992); upper Sandbian – lower Katian of British Columbia (Jin & Norford 1996); upper Sandbian – upper Katian of Sweden (Jaanusson 1962; Cocks 2005); lower Katian of the Southern Appalachians, eastern USA (Ulrich & Cooper 1942), Klamath Mountains, northern California, Jones Ridge in east-central Alaska and Yukon Territory, West Canada (Potter 1990c); upper Katian of eastern North Greenland (Rasmussen & Harper 2010); Rawtheyan of South China (Zhan & Cocks 1998) and finally reported from the Hirnantian? of the Northwest Territories, Canada (Potter 1990c). Potter & Boucot (1992) questionably reported the genus from the Hirnantian locality 79WG19. The current study does not confirm this report and in addition there is little evidence to support a Hirnantian age for locality 79WG19.

Bimuria (Bimuria) gilbertella Potter, 1991

Plate 4, figures 3–10

> 1991 *Bimuria gilbertella* n. sp. Potter, pp. 750–751, figs 2.20–2.29; figs 3.1–3.23.

Material. – One ventral valve and one indeterminable fragment from block 3 and 4 pairs of conjoined valves, 21 ventral and eight dorsal valves from blocks 4, 5 and 6. All from locality A-1230. This is part of Potter's (1991) original holotype and paratype material.

Description. – Shell strongly concavo-convex, transverse, subtriangular to subelliptical in outline, with obtuse cardinal angles. Ventral valve unevenly convex, almost bulbous medianly. Umbo located posteriorly. Maximum width about one-third of valve length from posterior margin. Anterior commissure rectimarginate. Maximum convexity near umbo. Ventral interarea anacline. Ventral umbo perforated in some specimens. Delthyrial angle wide, narrow pseudodeltidium present. Shallow, rounded median indentation extends from near umbo to anterior third of valve. Dorsal valve more concave posteriorly. Exterior median septum present. Dorsal interarea probably hypercline, but difficult to assess. Valves are smooth, lamellose in some specimens.

Ventral interior with large, widely spaced and divergent teeth. Dental plates not present. Ventral muscle field poorly impressed, but large and extended anteriorly. Dorsal interior with simple, probably undercut cardinal process. Chilidium present. Small alveolus just anterior to cardinal process; very short median septum, initially elevated, develops in front of alveolus. Two narrow-angled side septa are more elevated and extend for about 80% of valve length, flanking a bilobed bema.

Remarks. – Potter (1991) fully described the species of the assigned material. Two ventral valves and one dorsal valve are questionably assigned to this species, even though all three are part of Potter's (1991) original paratype material (paratypes of ventral valves USNM 413558 and USNM 413559 and dorsal valve USNM413562). These ventral valves are lamellose, more transverse and have a less bulbous umbonal convexity. Furthermore, the depression on the posterior half of the umbo is less accentuated on these valves than in the rest of the material and the anacline ventral interarea appears less inclined. However, there is no sign of geniculation. The dorsal valve has a more elevated bema than the rest of the material, but again there is no sign of geniculation. The cardinal process is simple, but appears undercut contrary to the diagnosis of this genus (Cocks & Rong 1989).

Genus *Craspedelia* Cooper, 1956

Type species. – *Craspedelia marginata* Cooper, 1956; from the Pratt Ferry Formation (upper Darriwilian) of Alabama, USA.

Biogeographic distribution. – Reported from the lower Sandbian– lower Katian of Girvan, Scotland (Williams 1962; Harper 1989) and Kazakhstan (Klenina *et al.* 1984); Sandbian of the Jones Ridge, east-central Alaska and Yukon Territory, western Canada and the Klamath Mountains, northern California (Potter 1991); upper Cautleyan–Rawtheyan of Sweden (Cocks 2005). Probably the youngest report is from east-central Alaska and Yukon Territory. There the genus occurs in late Katian or Hirnantian strata (Potter 1991). However, this study casts some doubt on the Hirnantion age of the material from the Jones Ridge area.

Craspedelia potterella n. sp.

Plate 4, figures 11–17

Derivation of name. – Named in honour of Dr. Alfred W. Potter, who provided a significant contribution to phylogenetic relationships of this family.

Holotype. – MGUH 29499 (Pl. 4, fig. 11).

Paratypes. – MGUH 29500 (Pl. 4, figs 12, 13), MGUH 29501 (Pl. 4, figs 14, 15) and MGUH 29502 (Pl. 5, figs 16, 17).

Material. – Three pairs of conjoined valves, 28 ventral and 24 dorsal valves from locality A-1230, blocks 4–6. One dorsal valve from locality 79WG19.

Diagnosis. – Shell medium-sized, with concavo-convex profile. Transverse outline with alae. Strongly geniculate in ventral direction. Smooth, lamellose ornamentation. Cardinal process undercut, with shaft almost beyond posterior margin. Long dorsal median septum, separating bilobed bema.

Description. – Shell medium-sized, with concavo-convex profile. Transverse, semilenticular outline, with almost alate cardinal angles. Maximum width at posterior margin. Strongly geniculate in ventral direction. Ventral valve unevenly convex with large, wide, umbo and rounded beak. Smooth, lamellose ornamentation which is especially marked in the anterior half where geniculation starts. Interarea relatively high, wide, almost orthocline. Dorsal valve is strongly concave, most concave near the posterior margin. Dorsal interarea probably hypercline, not well preserved. Chilidial plates or chilidium not present. Cardinal process undercut, with shaft almost beyond posterior margin with small alveolus developed anteriorly. Median septum developed anterior to alveolus. Median septum more elevated at its posterior end and extending to mid-valve. Socket ridges lie on valve floor, slightly more elevated towards the cardinal process and distally. Two narrow angled, high, side septa extend almost as far posteriorly as the median septum; towards the anterior margin they diverge laterally to form a highly elevated, bilobed bema. The broad septa have saw-like depressions anteriorly.

Remarks. – *Craspedelia potterella* differs from *C. intonsa* Potter 1991 from the Darriwilian Gregg Ranch unit in the Eastern Klamath Mountains, northern California, in lacking ridges and a sulcate anterior commissure, and from *Craspedelia* sp. 1 Potter, 1991 from the late Katian member 5 of the Horseshoe Gulch unit, also from the Eastern Klamath Mountains, in lacking a radial ornament. Furthermore it differs in outline from the middle Ordovician *C. tata* Popov, 1980 of Kazakhstan, *C. tata* being more triangular and from *C. marginata* Cooper, 1956 from the Darriwilian Pratt Ferry Formation of Alabama in having an undercut cardinal process. Differs from lower Katian *C.* cf. *gabata* Williams, 1962 (*in* Harper 1989) of the Myoch Formation, Girvan, Scotland, in having a more transverse outline.

See remarks in Potter (1991) for further discussion of the paratypes. Some of his paratypes of *Bimuria gilbertella* could also be ascribed to this species (see remarks for *B. gilbertella* above).

Family Leptellinidae Ulrich & Cooper, 1936
Subfamily Leptellininae Ulrich & Cooper, 1936

Genus *Leptellina* Ulrich & Cooper, 1936

Subgenus *Leptellina* (*Leptellina*) Ulrich & Cooper, 1936

Type species. – *L. tennesseensis* Ulrich & Cooper, 1936; from the Darriwilian Lenoir Formation, Tennessee, USA.

Biogeographic distribution. – Lower Sandbian of South China; upper Sandbian of Altai (Sennikov *et al.* 2008) and Tasmania (Laurie 1991); Sandbian–lower Katian of Myanmar (Cocks & Zhan 1998; Zhan & Cocks 1998), Ireland including Pomeroy (Mitchell 1977; Parkes 1992; Candela 2003) and Virginia, USA (Cooper 1956; Potter & Boucot 1992); middle Katian of Kazakhstan (Rong & Boucot 1998); Cautleyan of Pomeroy, Ireland (Mitchell 1977) and Rawtheyan of South China (Liu *et al.* 1983).

Leptellina (Leptellina) occidentalis Ulrich & Cooper, 1936

Plate 4, figures 18–24; Plate 5, 1–2

Material. – One ventral and two dorsal valves from locality A-1230, blocks 4–6.

Description. – Shell gently concavo-convex, transverse, semi-elliptical in outline, with acute, slightly alate cardinal angles. Anterior commissure rectimarginate. Ventral valve weakly to moderately convex, notably towards the anterior margin. Moderately high, weakly apsacline interarea. Small apical pseudodeltidium present. Dorsal interarea flat, low and catacline to hypercline. Chilidium present. Ornament unequally parvicostellate with accentuated costae relatively closely spaced.

Ventral interior with large teeth supported by small, widely divergent dental plates. Ventral muscle field strongly impressed, bilobed and extends anteriorly at about one-third of valve length from the posterior margin. Adductor scars commonly well confined in a small bilobed area enclosed by strong muscle bounding ridges. Dorsal interior with trifid cardinal process with fused socket ridges bases. Median septum

moderately elevated, separating two or three pairs of low ridges that become weaker laterally. Median septum extends almost to the anterior margin, but ends anteriorly in a divided, weakly elevated but variably developed subperipheral rim.

Remarks. – A platform-like structure with pronounced muscle bounding ridges parallel to, and just posterior of, the anterior margin, is typical for this genus. The current material is assigned to *L. occidentalis* based on the combination of an alate outline, the fused, trifid cardinal process, the lack of true side septa as well as the characteristic variation in the development of the subperipheral rim, that are also evident from Cooper's illustrations of the species (Cooper 1956, pl. 189C, figs 34–37). The same characters separate the current material from the type species, *L. tennesseensis* (see below).

Leptellina (Leptellina) tennesseensis Ulrich & Cooper, 1936

Plate 5, figures 3–8

Material. – Eight ventral and three dorsal valves, all from blocks 4–6 locality A-1230.

Description. – Shell strongly concavo-convex in profile and transverse, subelliptical in outline. Maximum width at hinge line or near posterior margin, with acute to right angled cardinal angles, possibly alate. Anterior commissure rectimarginate. Some specimens have a weak dorsal fold anteriorly. Ventral valve unevenly convex, more convex anteriorly, with a medium to high, wide apsacline interarea. Small apical pseudodeltidium present in some specimens. Dorsal valve moderately concave, with a low, wide anacline to catacline interarea. Chilidium absent. Unequally parvicostellate ornamentation, usually more marked on dorsal valves with 6–8 costellae per mm along anterior margin of mature specimens and with 7 accentuated costae originating at umbo and extending to the anterior margin.

Ventral interior with medium to large, broad teeth supported by dental plates that rise from valve floor. Ventral muscle field possibly bilobed and extends anteriorly to about one-fourth of valve length from the posterior margin. Adductor scars developed in a small confined area enclosed by muscle bounding ridges. Dorsal interior with large, bulbous trifid cardinal process. Socket ridges thick and crenulated; deep sockets. Dorsal median septum extends from the thick, swollen area anterior to cardinal process as thick elevation medially that merges with a strongly bilobed platform, that follows the shell outline. Platform disappears posteriorly just anterior to hinge line.

Remarks. – This species is assigned to *Leptellina* on the basis of the characteristic cardinal process, that is clearly different from that typical of, for instance, *Leptelloidea*. In addition the simple, fused cardinal process is very characteristic of the type species of *Leptellina*, as is the well developed, bilobed supperipheral rim which also distinguishes it from *L. occidentalis*, described above.

Genus *Leptelloidea* Jones, 1928

Type species. – *Plectambonites schmidti* var. *leptelloides* Bekker 1921; from the lower Sandbian Kukruse Formation, Estonia.

Biogeographic distribution. – Lower Sandbian of Estonia (Cooper 1956), South China (Chang 1983) and possibly from Southern Shan States, Myanmar (Cocks & Rong 1989).

Leptelloidea leptelloides Bekker, 1921

Plate 5, figures 9–14

Material. – One ventral and two dorsal valves from blocks 4–6 and possibly one dorsal fragment from block 3; all specimens are from locality A-1230.

Description. – Shell moderately concave, transverse, subelliptical in outline. Maximum width probably at hinge line. Anterior commissure rectimarginate. Dorsal valve strongly concave, with a high, wide catacline interarea. Chilidium absent. Unequally parvicostellate ornamentation with relatively closely spaced accentuated costellae that are intercepted by accentuated costae.

Dorsal interior with large bulbous cardinal process projecting posteriorly from a closed, convex chilidium. Socket ridges wing-like, bending at right angles in the anterior direction and not supported by valve floor, laterally. Dorsal median septum more like an elevated, thick ridge with a pair of side septa extending just anterior of cardinalia as short caluses on platform floor. Anteriorly the septum fuses with valve floor to form a bilobed elevated platform that continues laterally towards the cardinal extremities, extending for at least half of valve area. Muscle impressions faint.

Remarks. – This genus differs from *Leptellina* in having a more concave dorsal valve and having longer ventral diductor muscle impressions and cardinalia in the dorsal valve. Accentuated costellae are also more closely spaced in *Leptelloidea*. However, this species may be differentiated based on the very characteristic wing-like brachiophores and its raised, smaller platform (compared to *Leptellina*). This is the first

reported occurrence of the genus outside Estonia and the Southern Shan-States.

Family Leptestiidae Öpik, 1933

Genus *Diambonia* Cooper & Kindle, 1936

Type species. – *Plectambonites gibbosa* Winchell & Schuchert, 1895; from the upper Ordovician of Minnesota, USA.

Biogeographic distribution. – Known from the lower–upper Sandbian of Girvan, Scotland (Williams 1962) and Norway (Hansen 2008); upper Sandbian of Pomeroy (Mitchell 1977); middle–upper Katian of Girvan (Harper 2000), the Jones Ridge area (Ross & Dutro 1966), Pomeroy (Cautleyan, Mitchell 1977) and Quebec (Cocks 2005).

Diambonia discuneata Lamont, 1935

Plate 5, figures 15–20

1935 *Leangella discuneata* Lamont, p. 315, pl. 7, figs 17–19.
cf. 1962 *Diambonia* cf. *discuneata* (Lamont); Williams, p. 173, pl. XVI, figs 25–29.
1966 *Diambonia* sp. 1 Ross & Dutro, p. 15, pl. 2, figs 15, 16.
1970 *Diambonia discuneata* (Lamont); Cocks, p. 157, pl. 1, figs 9–10.
1977 *Leangella discuneata* Lamont; Mitchell, p. 78, pl. 15, figs 10–25.
1982 *Diambonia discuneata* (Lamont); Harper, fig. 14.
1989 *Diambonia discuneata* (Lamont); Harper, p. 108, pl. 18, figs 1–5, 8, 9, 12, 13.

Material. – One conjoined specimen and seven ventral valves from locality 79WG19. Further, one conjoined specimen from locality 79WG126.

Description. – Small, strongly concavoconvex with an alate rectangular outline. Maximum width at acute cardinal extremities. Ventral valve unevenly convex with a bulbous umbo that projects slightly posterior of hingeline. Strongest convexity towards anterior margin. Ventral interarea apsacline, slightly curved. Delthyrial angle wide. Dorsal valve essentially flat until at anterior margin where the shell rises vertically to become deeply concave in profile. Interarea hypercline. Ornament of five widely spaced, accentuated costae on an otherwise smooth exterior. Growth lines developed close to anterior margin.

Ventral interior with large, widely spaced and divergent teeth. Dental plates not present. Ventral muscle field strongly impressed, divided by strongly developed ventral median septum that elevates anteriorly. High muscle bounding ridges clearly separate adductor from diductor muscle fields.

Dorsal interior not known.

Remarks. – Although synonymized with *Leangella* and suppressed as a junior synonym by Mitchell (1977) and later by Cocks & Rong (1989), we have retained *Diambonia* as an independent genus. The above mentioned authors argued that in larger samples of the genus, the ventral median septum is variably developed to being almost absent in some specimens, whereas the exterior ornament is supposedly identical between *Leangella* and *Diambonia*. Thus, they have argued for the rejection of the name *Diambonia*. However, material of the type species of *Diambonia*, *D. discuneata*, from the type locality in the Girvan district consistently possess a ventral median septum. Thus, we have chosen to retain *Diambonia* as an independent genus.

The occurrence of *Diambonia* in the Farewell Terrane further strengthens the biogeographic affinity of this exotic terrane with the Midland Valley Terrane, but also the Jones Ridge area as it is here synonymized with *Diambonia* sp. 1 of Ross & Dutro (1966).

Family Xenambonitidae Cooper, 1956

Subfamily Xenambonitinae Cooper, 1956

Genus *Xenambonites* Cooper, 1956

Type species. – *Xenambonites undosus* Cooper 1956; from the Pratt Ferry Formation (upper Darriwilian) of Alabama, USA.

Biogeographic distribution. – Reported from the Jones Ridge area, east-central Alaska (Ross & Dutro 1966), in rocks that are probably contemporaneous with the White Mountain material. In addition, the genus is known from the lower Sandbian of Girvan (Williams 1962)

Xenambonites revelatus Williams, 1962

Plate 5, figures 21–24; Plate 6, figure 1

   1962 *Xenambonites revelatus* sp. nov. Williams, p. 191, pl. XVIII, figs 21–23.

   1966 *Xenambonites* cf. *revelatus* Williams; Ross & Dutro, p. 16, pl. 2, figs 21–26.

Material. – 47 ventral and 57 dorsal valves. All from locality A-1230, blocks 4–6.

Description. – Shell medium to large, strongly concavo-convex with lateral profile geniculate in ventral direction. Transverse, triangular outline. Maximum width at hinge line. Cardinal angles acute. Valves about 33% as long as wide. Ventral valve unevenly convex with a fold medially. Interarea moderately high, wide, and apsacline. Delthyrium wide and open. Dorsal valve concave with pronounced, wide sulcus developed towards the anterior margin. Small dorsal fold variably developed medially from within the umbonal zone to the anterior margin. Interarea low, wide and cataline. Chilidium covers about half of notothyrium. Finely unequally parvicostellate ornamentation, usually with 3–5 accentuated ribs and 12–14 parvicostellae per mm along anterior margin of mature specimens. Rugae, or concentric wrinkles, are developed anteriorly and laterally on some specimens.

Ventral interior with large teeth supported by divergent vestigal dental plates. Small ventral muscle scars elevated anteriorly. Platform surrounding muscle scars extends for 60% of valve length. Ribbed ridge dividing diductor muscle scars. Dorsal interior with undercut cardinal process. Strong socket ridges confined within small bilobed elevated bema. Weak ridges support the bema at an angle anterior to the bema. Dorsal median septum lacking.

Remarks. – The present specimens are assigned to *X. revelatus* on the basis of the finely parvicostellate ornamentation, rather than to the slightly older *X. undosus* Cooper from Alabama, that possesses a coarser ornament. The deep elongate sockets between the socket ridges and the hinge-line, was used by Ross & Dutro (1966) as a diagnostic character for this species. Moreover, the concentric rugae confirm to the observations of Ross & Dutro (1966) on the material of this species from Jones Ridge. They only questionably assigned their limited number of specimens to *X.* cf. *revelatus*, since wrinkles were not seen on the type material from Girvan. Their specimens also possessed a larger, dorsal median fold. However, with this new material from the White Mountain area, it is clear that both the concentric ornamentation and the dorsal median fold vary strongly within the population. Thus both the material from Jones Ridge and this material from the White Mountain area should be assigned to *X. revelatus* Williams, 1962.

Subfamily Aegiromeninae Havlíček, 1961

New genus aff. *Aegiria*

Plate 6, figures 2–7

Material. – One ventral and five dorsal fragments from locality 79WG19.

Description. – Shell elongately oval with a weakly concavo-convex to plano-convex profile. Ventral interarea apsacline, curved laterally in anterior direction. Ornamentation of simple, thick costae, lamellose towards anterior margin.

Ventral interior with muscle field developed on raised platform. Dental plates divergent, forming an acute angle. Dorsal interior with divergent dental sockets; inner socket ridges short. Median septum elevated, flanked laterally by two to three pairs of lateral ridges broadly describing a small raised bema-like structure. Median septum transects anterior part of bema. Anterior margin developed as thick rim.

Remarks. – This material is here assigned to a new genus that has some close affinities with *Aegiria* based on the relatively transverse bema that is not bilobed anteriorly; the short septum that transects the anterior part of the bema and the lack of clearly defined side septa. The material illustrated here may represent two different species, one with a bema and one with a platform resembling that of *Anoptambonites*. However, this genus appears strongly lamellose towards the anterior margin, unlike *Anoptambonites*. Neither is the manifestation of a bema seen in *Anoptambonites*.

Family Hesperomenidae Cooper, 1956

Genus *Anoptambonites* Williams, 1962

Type species. – *Leptaena grayae* Davidson, 1883; from the lower Katian Craighead Limestone of Girvan, Scotland.

Biogeographic distribution. – Reported from the lower Sandbian of Giuzhou, South China (Xu *et al.* 1974) and Newfoundland (McKerrow & Cocks 1981); In Girvan the genus is reported from the lower Sandbian – upper Katian (Williams 1962; Harper 1989); upper Sandbian of Altai (Sennikov *et al.* 2008); Sandbian–middle Katian of the Chinghiz and Chu-Ili plates, Kazakhstan (Cocks & Rong 1989; Nikitin *et al.* 2006; Popov & Cocks 2006; Popov *et al.* 2000, 2002), New South Wales, Australia (Percival 1991); Sandbian–Rawtheyan of the Klamath Mountains, northern

California (Potter 1990b); Pusgillian–Rawtheyan of Sweden (Sheehan 1973; Cocks 2005); lower–middle Katian of Tien-Shan (Cocks 2005); lower Katian and Cautleyan of Pomeroy, Ireland (Mitchell 1977; Candela 2003); middle–upper Katian of the Prague Basin, Czech Republic (Havlíček 1982), Taimyr, Russia (Cocks & Modzalevskaya 1997), Wales (Cautleyan–Rawtheyan) (Mitchell 1977; Price 1981); uppermost Katian (Rawtheyan) of Estonia (Rõõmusoks 1963), Maine, USA (Neuman 1994), Oslo Region, Zhejiang, South China (Cocks & Rong 1989). Moreover, the genus is known from beds of probably Katian age in the Gornoi Altai Mountains (Severgina 1984) and from the Jones Ridge area, east-central Alaska (Ross & Dutro 1966) in beds of contemporaneous age with the current material.

Anoptambonites grayae (Davidson, 1883)

Plate 6, figures 8–14

1883 *Leptaena grayae* Davidson, p. 171, pl. 12, figs 23–25.
1883 *Leptaena llandeiloensis* Davidson, pp. 171–172, pl. 12, figs 27–29; *non* fig. 26.
1917 *Leptella grayae* (Davidson); Reed, p. 873, pl. 13, figs 10–13, 15–17.
1928 *Leptelloidea grayae* (Davidson); Jones, pp. 489–490.
1962 *Anoptambonites grayae* (Davidson); Williams, p. 171, pl. XVI, figs 11–14, 17.
1966 *Anoptambonites* cf. *grayae* (Davidson); Ross & Dutro, p. 12, pl. 2, figs 1, 3, 5, 7, 9.
1989 *Anoptambonites* aff. *grayae* (Davidson); Harper, p. 105, pl. 18, fig. 17.

Material. – Four ventral and one dorsal valve from block 3 and two pairs of conjoined valves; 51 ventral and 68 dorsal valves from blocks 4, 5 and 6. All specimens are from locality A-1230.

Description. – Shell moderately to strongly concavo-convex with transverse to subelliptical in outline. Almost as long as wide, however cardinal angles not preserved in the material. Maximum width just anterior of hinge line. Anterior commissure possibly rectimarginate, though the dorsal valve possesses a relatively pronounced sulcus, or depression, developing from about mid-valve. Ventral valve, unevenly convex; maximum convexity at mid-valve. High, narrow cataclilne to weakly apsacline interarea. Delthyrium open. Dorsal valve strongly concave; maximum concavity at anterior margin. Median depression develops anteriorly until just before the anterior margin, where the valve bends in dorsal direction. Interarea strongly anacline; large chilidium present.

Unequally parvicostellate ornamentation. One or two orders of accentuated costellae originate between umbonal area and mid-valve, increasing in numbers in mature specimens. Concentric growth lines variably developed, especially towards the anterior margin.

Ventral interior with large teeth supported by short dental plates. Ventral muscle field bilobed, moderately to strongly impressed extending anteriorly at about one-sixth of valve length. Ventral myophgram may be present in one specimen, however this may be due to preservation (see Pl. 5, fig. 22). Dorsal median septum with undercut cardinal process and divergent socket ridges. Alveolus present between socket ridges and median septum. Median septum more elevated anteriorly and merges at about two-third of valve length from posterior margin with platform rim. Platform forming one lobe unlike bilobed platform of *Leptellina* and *Leptelloidea*. Muscle scars faint.

Remarks. – Dorsal interior resembles *Leptellina* but differs in having a semicircular, non-elevated platform although Williams's type species of *A. grayae* has a bilobed anterior margin of the platform (Williams 1962; pl. 16, fig. 13). This character is not present in the Alaskan species. However, this character is known to show great variation in other species of *Anoptambonites* (Potter 1990b). The current material is thus assigned to *A. grayae* primarily on the basis of the alveolus between the socket ridges and the dorsal median septum – a character that is consistently absent in for example *A. myriomorpheus* from the Eastern Klamath Mountains, where the median septum is fused directly with the socket ridges (Potter 1990b). Moreover, the current study indicates that this species is identical to *A.* cf. *grayae* described and illustrated by Ross & Dutro (1966) from Jones Ridge, east-central Alaska. Contrary, however, to Potter (1990b) it is unlikely that the Jones Ridge material is conspecific with *Anoptambonites* sp. 1 from the Klamath Mountains.

Anoptambonites pulchra (Cooper 1956) new combination

Plate 6, figures 15–20

1956 *Leptellina pulchra* Cooper, new species, p. 753, pl. 189, A, figs. 1–23; pl. 195, C, figs. 9–16; pl. 219, G, fig. 14.

Material. – One pair of conjoined valves, 28 ventral and 40 dorsal valves. All from locality A-1230, blocks 4–6.

Description. – Shell strongly concavo-convex to ventrally resupinate; outline transverse to subelliptical, about 60% as long as wide. Maximum width at hinge line or near posterior margin, with acute to right angled cardinal angles. Some specimens with weak dorsal fold anteriorly. Ventral valve evenly convex, with a medium to high, wide apsacline interarea. Small apical pseudodeltidium present in some specimens. Dorsal valve strongly concave, with a low, wide, anacline to cataline interarea. Chilidium absent. Unequally parvicostellate ornamentation, usually stronger on dorsal valves; 6–8 costellae per mm along anterior margin of mature specimens and 7 accentuated costae originating at umbo, extending to the anterior margin. Some specimens lamellose.

Ventral interior with small, divergent dental plates and weak elongate to bilobed muscle scar impressions. Dorsal interior with undercut cardinal process, brachiophores with divergent socket ridges merging to form a plate. In front of alveolus, the dorsal median septum becomes more elevated anteriorly and merges with bilobed platform elevated on all sides. Commonly the median septum is weakly bent towards the left when viewed perpendicularly. Muscle scar impressions faint. Transmuscle septa occasionally developed.

Remarks. – This species is assigned to the Hesperomenidae based on the resemblance of the dorsal interiors to those of *Anoptambonites*. Ornamentation, however, is very different from the known species within this genus. Almost certainly con-specific with *Leptellina pulchra* Cooper, 1956 based on ornamentation, outline and the presence of a trifid, undercut cardinal process. Cooper's (1956) other species of *Leptellina* also have a trifid cardinal processes, but they are not undercut, a feature, that is characteristic of the Hesperomenidae (Cocks & Rong 1989). Cooper's original material was retained within *Leptellina* by Cocks & Rong (1989, p. 104). With regard to the outline and ornamentation of the ventral valve, this species resembles *Leangella*. However, the current species differs from *Leangella* in lacking a large ventral median septum.

Anoptambonites sp.

Plate 6, figures 21–22

Material. – One ventral and three dorsal valves from locality 79WG126.

Description. – Similar to *A. grayae*, but elongately oval (longer than wide) and with broader costellae. Lacks depression medially. Socket ridges less divergent.

Dorsal median septum fused to the socket ridges, i.e. no alveolus present. Platform also more elongately oval than in *A. grayae*.

Remarks. – This material may be conspecific with the older species described above, but for the reasons noted above this is unlikely. Like *A. myriomorpheus* from the Eastern Klamath Terrane (Potter, 1990b), the dorsal median septum is fused to the socket ridges but these appear less divergent. It differs from *Anoptambonites* sp. 1 from the Klamath Mountains in outline and in the disposition of the cardinalia (Potter 1990b). With only one specimen collected, it is left under open nomenclature.

Genus *Kassinella* Borissiak, 1956

Subgenus *Kassinella* (*Trimurellina*) Mitchell, 1977

Type species. – *Trimurellina superba* Mitchell, 1977; from the middle–upper Katian Killey Bridge Formation of Pomeroy, Northern Ireland.

Biogeographic distribution. – Middle–upper Katian of Pomeroy, Northern Ireland (Mitchell 1977) and the Hirnantian of Wales (Williams & Wright 1981).

Kassinella (*Trimurellina?*) sp.

Plate 6, figures 23–24; Plate 7, figures 1–5

Material. – Five ventral and nine dorsal valves. All from locality 79WG19.

Description. – Shell strongly concavo-convex, elongately oval in outline. Maximum width at cardinal angles. Anterior commissure rectimarginate. Ventral valve unevenly convex with strongest convexity near the anterior margin. Interarea high and strongly apsacline. Delthyrium open, high and wide. Dorsal valve weakly concave posteromedially, strongly concave towards the anterior commissure. Unequally parvicostellate ornamentation. Interarea wide and anacline; chilidium large. Concentric ornamentation variably developed, especially towards the anterior margin.

Ventral interior with short, recessive dental plates (teeth not seen). Ventral muscle field strongly impressed. Elongate, thin scars parallel to slightly oblique, separated by a thin, sharp ventral median septum. Vascula media preserved on one specimen. Dorsal interior with trifid cardinal process, possibly undercut, but the poor preservational state does not permit accurate observations. Brachiophores

divergent, more elevated laterally. Dorsal median septum more elevated anteriorly, merging at about one-third of valve length from anterior margin with non-divided peripheral rim; median septum thicker and hollow near anterior margin where it intersects platform at half of valve length. Platform slightly elevated and supported by lateral ridges developed about half way between cardinal angles and median septum. Muscle scar impressions faint.

Remarks. – The specimens are identified as *Kassinella* (*Trimurellina?*) rather than *K.* (*Kassinella*) on the basis of the median septum extending anterior to the platform but restricted anteriorly by the peripheral rim. It may be confused with the subgenus *Leangella* (*L.*) due to the possession of a subperipheral rim. However, the possession of the dorsal septum that extend anteriorly beyond the platform rim and an evenly parvicostellate ornamentation indicates assignment of these specimens to *Kassinella* (*Trimurellina*).

Genus *Sampo* Öpik, 1933

Type species. – *S. hiiuensis* Öpik 1933; from the Vormsi Stage (middle Katian) of Estonia.

Biogeographic distribution. – Lower Katian of Newfoundland (McKerrow & Cocks 1981); middle Katian of Estonia and Lithuania (Öpik 1933; Paškevičius 1994); Cautleyan of Pomeroy, Northern Ireland (Mitchell 1977); upper Cautleyan–Hirnantian of northern Wales (Hiller 1980) and the Hirnantian of southern Norway (Cocks 1988).

Sampo? sp.

Plate 7, figures 6–7

Material. – One ventral valve from locality 79WG126.

Description. – Ventral valve large, ellipsoidal to rectangular in outline. Maximum width at hinge line. Interarea strongly convex and strongly apsacline to almost orthocline. Ornament consisting of 26 simple costae. Interior with prominent, broad teeth; teeth perforated by deep, narrow grooves. Dental plates not seen. Diductor muscle field bilobed, strongly impressed and enclosed by muscle-bounding ridges.

Remarks. – This species is questionably assigned to *Sampo*, as no dorsal valves have been recovered within the material.

Plectambonitoidea gen. *et* sp. indet. 1

Plate 7, figures 8–9

Material. – One ventral valve from locality 79WG126.

Description. – Fragment of convex ventral valve with only a few accentuated costae. Interarea strongly apsacline with prominent teeth. Interior with diductor muscle scars developed as long, narrow and deep impressions bounded by ridges on either side.

Remarks. – Only one fragment was retrieved, that is not possible to assign at the genus level.

Order Billingsellida Schuchert, 1893

Suborder Clitambonitidina Öpik, 1934

Superfamily Clitambonitoidea Winchell & Schuchert, 1893

Family Clitambonitidae Winchell & Schuchert, 1893

Genus *Atelelasma* Cooper, 1956

Type species. – *A. perfectum* Cooper 1956; from the upper Darriwilian of Tennessee, USA.

Biogeographic distribution. – Reported from the lower Sandbian of Virginia and Tennessee, eastern USA (Cooper 1956), Siberia (Kanygin *et al.* 1989) and Ireland (Parkes 1992).

***Atelelasma*? sp**

Plate 7, figures 10–11

Material. – One fragment from blocks 4–6, locality A-1230.

Description. – Valve exterior multicostellate with fine ribs and strongly lamellose near the anterior margin. Aditicules seen on costellae.

Remarks. – The imbricate shell margin resembles that seen in clitambonitoids and, in particular, in *Atelelasma* – one of the few clitambonitoid genera known from Laurentia.

Order Orthotetida Waagen, 1884

Suborder Triplesiidina Moore, 1952

Superfamily Triplesioidea Schuchert, 1913

Family Triplesiidae Schuchert, 1913

Genus *Triplesia* Hall, 1859

Type species. – *Atrypa extans* Emmons 1842; from the Sandbian of New York State, USA.

Biogeographic distribution. – Lower Sandbian of Armorica (Botquelen & Mélou 2007) and Girvan (Williams 1962); Sandbian of Chu-Ili, Kazakhstan (Popov *et al.* 2002); lower Sandbian and Katian of Montagne Noire, France (Hammann *et al.* 1982); Sandbian–lower Katian of Iberia (Villas 1992; Botquelen & Mélou 2007); Katian of Newfoundland (McKerrow & Cocks 1981); lower Katian of Altai (Severgina 1978; Kulkov & Severgina 1989), northeastern Kazakhstan (Nikitin *et al.* 2003), Morocco and Perunica (Villas 1992) and the Oslo Region (Harper *et al.* 1984; Hansen 2008); Pusgillian of Girvan (Harper 2006); Cautleyan of Pomeroy, Ireland (Wright 1964); Rawtheyan of Belgium (Sheehan 1987), Girvan (Harper 2006) and Taimyr (Cocks & Modzalevskaya 1997); Rawtheyan–middle Hirnantian of South China (Rong *et al.* 2002; Zhan *et al.* 2002); Hirnantian of Kolyma and the Oslo Region (Cocks 1988).

***Triplesia* sp.**

Plate 7, figures 12–13

Material. – Thirteen dorsal valves. All from locality A-1230, blocks 4–6.

Description. – Dorsal valve evenly convex, with strongest convexity at or near hinge line. Valves smooth. Interior with long, protruding forked, bifid cardinal process that extends well beyond posterior margin, keeled proximally with cowl. At hinge line it merges with brachiophores that are curved and bent in a posterior direction laterally beyond the hinge line. No other internal features visible.

Remarks. – Though with certainty assigned to *Triplesia*, the current fragments cannot be identified at the species level. All the material fragmented, and the outline is impossible to assess. Probably subrounded, with obtuse cardinal extremities.

Genus *Grammoplecia* Wright & Jaanusson, 1993

Type species. – *G. triplesioides* Wright & Jaanusson, 1993; from the upper Katian Boda Limestone, Siljan, Sweden.

Biogeographic distribution. – Sandbian–lower Katian of the Chu-Ili Terrane (Popov *et al.* 2000, 2002; Popov & Cocks 2006); upper Cautleyen–Rawtheyan of central Sweden (Cocks 2005).

Grammoplecia? sp.

Plate 7, figure 14

Material. – Two fragments from locality 79WG19 and one fragment from locality 79WG126.

Description. – Capillate ornament intersected by closely spaced filae.

Remarks. – Though only a small fragment is available, the characteristic ornamentation closely resembles that of *Grammoplecia*.

Class Rhynchonellata Williams *et al.*, 1996
Order Protorthida Schuchert & Cooper, 1931
Superfamily Skenidioidea Kozłowski, 1929
Family Skenidiidae Kozłowski, 1929

Genus *Skenidioides* Schuchert & Cooper, 1931

Type species. – *Skenidioides billingsi* Schuchert & Cooper, 1931; from the lower Katian of Québec, eastern Canada.

Biogeographic distribution. – Reported from the Sandbian of Altai (Sennikov *et al.* 2008); the Chu-Ili Range, Kazakhstan (Popov *et al.* 2002), Klamath Mountains, northern California (Potter 1990b), Northwest Territories, Canada, Oklahoma, Minnesota, Wisconsin, Illinois, Virginia and Maine (Cooper 1956; Potter & Blodgett 1992), South China (Xu *et al.* 1974; Rong *et al.* 1999); Sandbian–lower Katian of Autralia (Laurie 1991; Percival 1991); Sandbian–middle Katian of the Midland Valley Terrane (Williams 1962; Mitchell 1977; Harper 2000, 2006; Candela 2003; Candela & Harper 2010) and Myanmar (Cocks & Zhan 1998); upper Sandbian – lower Katian of the Oslo Region (Hansen 2008); upper Sandbian–Hirnantian of Avalonia (Williams 1963; Sheehan 1987; Cocks 1988; Parkes 1992); Katian of North America (Eastern Klamath

and Sierra terranes of northern California, Northwest Territories, Canada, Nevada, Maine, Québec and Newfoundland) (Potter & Blodgett 1992); lower Katian of northern Kazakhstan (Nikitin *et al.* 2003), South China (Rong *et al.* 1999) and North China (Rong *et al.* 1999); middle–upper Katian and Spain (Villas 1985); Rawtheyan of Maine, USA (Neuman 1994), Taimyr (Cocks & Modzalevskaya 1997) and Zhejiang, South China (Zhan *et al.* 2002) and the Hirnantian of the Oslo Region (Baarli & Harper 1986; Baarli 1995).

Skenidioides multifarius Potter, 1990b

Plate 7, figures 15–20

> 1990b *Skenidioides multifarius* n. sp. Potter, p. 16, pl. 3, figs 42–59.
> 1990b *Skenidioides* cf. *multifarius* n. sp. Potter, pl. 3, figs 60–66; pl. 4, figs 1–5.
> 1990b *Skenidioides* aff. *multifarius* n. sp. Potter, p. 19, pl. 4, figs 6–24.

Material. – One pair of conjoined valves, 70 ventral and 19 dorsal valves from locality A-1230, blocks 4–6. Two pairs of conjoined valves, 1 ventral valve and 2 dorsal valves are from locality 79WG19 and, lastly, 1 ventral and 2 dorsal valves from locality 79WG126.

Description. – Shell strongly ventri-biconvex, transverse to subtriangular in outline. Maximum width at hinge line. Anterior commissure sulcate. Ventral valve un-evenly convex, almost triangular or subpyramidal with marked umbo. Apex is positioned more posterior than hinge-line. Ventral interarea high, weakly apsacline. Delthyrium high, wide. Dorsal valve weakly convex with sulcus stronger at the anterior margin. Dorsal interarea wide, strongly anacline. Notothyrium wide. Simple costate to weakly parvicostellate ornamentation, usually with 15 costae on ventral valve and 13 costae on the dorsal valve.

Ventral interior with large deltidiodont teeth supported by dental plates that fuse medianly to form a spondylium simplex. The spondylium simplex is supported, at a low angle, by a very small ventral median septum. Dorsal interior with marked, blade-like cardinal process. Notothyrium open. Deep sockets are defined anteriorly by relatively long, moderately divergent brachiophores, supported by plates that fuse to form the cruralium converging onto linear, dorsal median septum that extends to anterior margin. Some specimens have a pair of low side septa or muscle bounding ridges that originate on the valve floor posteriorly and converge to the median septum near the anterior margin.

Remarks. – Whereas the material from locality A-1230 resembles *S.* cf. *multifarius* of Potter (1990b), the material from locality 79WG19 resembles Potter's *S.* aff. *multifarius*. In the current study the specimens of *Skenidioides* from the two localities are described as a single species.

Genus *Replicoskenidioides* Potter, 1990b

Type species. – *Replicoskenidioides rodneygreggi* Potter, 1990; from the Katian limestone members of the Horseshoe Gulch unit, Klamath Mountians, northern California, USA.

Biogeographic distribution. – Hitherto only known from the Klamath Mountains, northern California (Potter 1990b).

Replicoskenidioides sp.

Plate 7, figures 21–24

Material. – One conjoined pair and two dorsal valves from locality 79WG19.

Description. – Shell very small, rectangular in outline, with convexoplanar to ventri-biconvex profile. Maximum width at hinge line. Anterior commissure rectimarginate. Ventral valve evenly convex, subpyramidal with marked umbo. Umbo is positioned more posterior than hinge line, but not as posterior as in *Skenidioides*. Ventral interarea is very high, strongly apsacline. Delthyrium, high and very wide. Dorsal valve flat. Dorsal interarea wide, strongly anacline. Notothyrium wide. Ornamentation consists of 10 thick, rounded costae, widely spaced. Median costae lacking on ventral valve but low, broad costae developed medially on dorsal valve.

  Ventral interior lacks spondylium. Dorsal interior with swollen cardinal process continuous anteriorly with high, elevated thick median septum. Notothyrium open. Deep sockets are marked anteriorly by low lying, thick and long brachiophores, highly divergent, almost parallel to hinge line.

Remarks. – This material is assigned to *Replicoskenidioides* based on its similarity with the Klamath material of Potter (1990b). However, the dorsal sulcus is very shallow, if present at all, and the ornamentation consists of apparently coarser costae than those seen in the type species. Therefore, this could be a new species, or, a juvenile specimen where plication has not yet been developed at the anterior commissure. Not enough material has been found to determine which of the two possibilities is most likely.

Skenidiid gen. n. et sp. indet.

Plate 8, figures 1–2

Material. – One dorsal valve from locality A-1230, blocks 4–6.

Description. – Shell small, convex and transversely lobate, about 55% as long as wide; cardinal angles alate; maximum width at hinge line; anterior commissure sulcate; maximum convexity near anterior margin. Interarea very high, weakly anacline; notothyrium open, wide. Ornamentation of simple, accentuated costae to weakly parvicostellate; ribs are triangular in profile and widely spaced, numbering 22–24 costellae at anterior margin.

  Dorsal interior with high, sharp, narrowly spaced brachiophores; dental sockets deepest posteriorly. Notothyrial platform deep, narrow; cardinal process strong, bladelike, merging anteriorly with dorsal median septum more elevated towards the anterior margin.

Remarks. – This is probably a new genus on the basis of its distinctive shape and ornament. However, the paucity of the material inhibits the erection of a new genus with confidence.

Order Orthida Schuchert & Cooper, 1932

Suborder Orthidina Schuchert & Cooper, 1932

Superfamily Plectorthoidea Schuchert & LeVene, 1929

Family Plectorthidae Schuchert & LeVene, 1929

Subfamily Plectorthinae Schuchert & LeVene, 1929

Genus *Plectorthis* Hall & Clarke, 1892

Type species. – *Orthis plicatella* Hall, 1847; from the middle Katian of the USA.

Biogeographic distribution. – Recorded from the lower Sandbian of New South Wales, Australia (Percival 1991); lower Sandbian–lower Katian of the Chu-Ili Terrane, Kazakhstan (Popov *et al.* 2002; Popov & Cocks 2006) and Gorny Altai, Russia (Kulkov & Severgina 1989); lower Sandbian – middle Katian of the Midland Valley Terrane (Williams 1962; Mitchell 1977; Candela 2003); Katian of eastern USA (Cocks & Modzalevskaya 1997; Patzkowsky & Holland 1997); middle–upper

Katian of Taimyr, northern Russia (Cocks & Modza-levskaya 1997) and finally the Rawtheyan of South China (Rong *et al.* 1999; Zhan *et al.* 2002).

Plectorthis sp. 1

Plate 8, figures 3–6

Material. – 11 ventral, two dorsal valves and five fragments. All are from blocks 4, 5 and 6 at locality A-1230.

Description. – Shell large, plano-convex to weakly ventri-biconvex; outline difficult to determine due to the fragmented state of the material, but probably subrectangular. Cardinal extremities not preserved. Anterior commissure weakly sulcate. Ventral valve evenly convex; interarea probably very short, curved and apsacline; delthyrium moderately high and wide. Dorsal valve flat to weakly convex, with sulcus most pronounced at the anterior margin. Dorsal interarea high, wide and strongly anacline; notothyrium open, large and wide. Ornamentation consisting of simple, widely spaced, large costae widening and more rounded at the anterior margin. One or two growth lines usually developed near the anterior margin.

Ventral interior with large deltidiodont teeth supported by divergent dental plates that fuse on valve floor to join muscle scars which are separated by a thick ridge – the adductor muscle track – elevated anteriorly on callus. Ventral median septum absent. Dorsal interior with deep sockets and widely divergent brachiophores that surround a small blade-like cardinal process developed on notothyrial platform, not elevated from valve floor. Short thickening extending about one-third of valve length, anterior of notothyrial platform; dorsal median septum lacking.

Remarks. – The exopunctate ornamentation and complex cardinalia consisting of fulcral plates supporting a platform with a thin, blade-like cardinal process, closely resembles that of *Plectorthis*. However, better preserved material is necessary to identify these specimens at the species level.

Plectorthis sp. 2

Pl. 8, figures 7–10

Material. – Seven ventral valves, nine dorsal valves and three fragments. All are from blocks 4, 5 and 6 at locality A-1230.

Description. – Shell subcircular, medium sized, plano-convex to weakly ventribiconvex in profile. Cardinal extremities obtuse, maximum width at mid-valve length. Anterior commissure rectimarginate to weakly sulcate. Ventral valve evenly convex; interarea wide, medium high, straight and apsacline; delthyrium high and wide. Dorsal valve flat to weakly convex, with sulcus most pronounced at the anterior margin. Dorsal interarea flat, wide and anacline; notothyrium open, large and wide. Ornamentation consists of simple triangular shaped costae that flatten and widen towards the anterior margin. The costae are relatively widely spaced, but not as pronounced and rounded as in *Plectorthis* sp. 1. One or two growth lines are often developed near the anterior margin. Maximum number of costae is 19.

Ventral interior with large deltidiodont teeth supported by relatively divergent dental plates that join valve floor to form a callus; diductor muscle scars parallel, long, slender and widely spaced, developed on callus. Diductor muscle scar impressions separated by a thickening of the callus where an elongate, narrow adductor muscle scar is developed.

Dorsal interior with deep dental sockets and widely divergent brachiophores that meet perpendicularly in notothyrium and confine a blade-like cardinal process on notothyrial platform, slightly elevated from valve floor.

Remarks. – *Plectorthis* sp. 2 is differentiated from *Plectorthis* sp. 1 in possessing longer, less divergent brachiophores and flatter, broader costae. Further, the callus is better developed in *P.* sp. 1.

Genus *Desmorthis* Ulrich & Cooper, 1936

Type species. – *Desmorthis nevadensis* Ulrich & Cooper, 1936; from the Darriwilian (Whiterockian, Middle Ordovician) Poconip Group, Nevada, USA.

Biogeographic distribution. – Previously only known from Nevada. This is the first record outside Laurentia.

Desmorthis? sp.

Plate 8, figures 11–12

Material. – One dorsal fragment from blocks 4–6, locality A-1230.

Description. – Dorsal valve medium sized, weakly convex with relatively high, though probably narrow, anacline interarea. Notothyrium large, wide and open. Ornamentation of simple, closely spaced costae. Dorsal interior with relatively narrow, high brachiophores flanking long, blade-like cardinal process; dental sockets narrow, relatively deep. Notothyrial platform extends anterior of brachiophores and thickens anteriorly as a low dorsal median septum. Lateral ridges and adductor muscle depressions present.

Remarks. – If correctly identified, this suggests that some of the fauna is of late Darriwilian age, or, alternatively the range of this genus should be extended upwards.

Genus *Doleroides* Cooper, 1930

Type species. – *Orthis gibbosa* Billings, 1857; from the Sandbian of Minnesota, USA.

Biogeographic distribution. – Reported from the lower Sandbian of the Mackenzie Mountains and Ontario, Canada and eastern USA (Cooper 1956; Ludvigsen 1975) and Siberia (Andreeva & Nikiforova 1955); upper Sandbian of Oklahoma (Alberstadt 1973); upper Sandbian – lower Katian of the Midland Valley Terrane (Mitchell 1977) and New South Wales (Percival 1991, 2009). In addition, it is reported from lower Sandbian beds of the northern Kuskokwim Mountains (Nixon Fork Subterrane), west-central Alaska (Measures *et al.* 1992; Rohr *et al.* 1992).

***Doleroides* sp.**

Plate 8, figures 13–17

Material. – One ventral and one dorsal valve from blocks 4–6, locality A-1230.

Description. – Shell medium to large, with ventribiconvex profile and subcircular outline. Cardinal angles obtuse, maximum width at mid-valve length. Anterior commissure weakly sulcate. Ventral valve evenly convex with umbo posterior of hinge line. Interarea strongly apsacline, strongly curved, wide and high. Delthyrium, open, high and wide. Dorsal valve weakly convex, with weak sulcus. Dorsal interarea high, wide and strongly anacline; notothyrium open and very wide. Ornamentation multicostellate consisting of even sized and spaced first and second order costellae that originate in the umbonal region. Costellae are often hollow and therefore appear punctured. Minute holes developed on hollow costae.

Ventral interior with low, wide teeth supported by large, narrow dental plates that merge onto valve floor to form medially thickened callus. Diductor muscle impressions are elongate, divergent anteriorly. Dorsal interior with a differentiated cardinal process, posteriorly low and blade-like but anteriorly becoming elevated. Brachiophores are relatively long, divergent and

at right-angle. Brachiophores supported by brachiophore plates that fuse onto valve floor to form a low lying notothyrial platform. This may continue anteriorly as a short thickening of the valve floor medially, but not as a median septum.

Remarks. – This form may be conspecific with *Doleroides* n. sp. aff. *D. panna* (Andreeva), which is reported below. That species is further known from the northern Kuskokwim Mountains, in the more northerly part of the Nixon Fork Subterrane (Measures *et al.* 1992; Rohr *et al.* 1992). Here the two species are provisionally retained as the transverse outline of *Doleroides* n. sp. aff. *D. panna* differs from that of *Doleroides* sp.

***Doleroides* n. sp. aff. *D. panna* (Andreeva)**

Plate 8, figures 18–20

> aff. 1955 *Mimella panna* Andreeva sp. nov., p. 65, pl. XXVII, figs 1–6.
> 1992 *Doleroides* n. sp. aff. *D. panna* (Andreeva); Measures, Rohr & Blodgett, fig. 8A, B.
> 1992 *Doleroides* n. sp. aff. *D. panna* (Andreeva); Rohr, Dutro & Blodgett, figs 8.14–8.24.

Material. – Four ventral valves from blocks 4–6, locality A-1230.

Description. – Ventral valve medium-sized, convex and semi-oval to rectangular in outline. Cardinal angles acute to at right angle. Maximum width within one-third of the anterior margin; anterior commissure probably rectimarginate. Ventral valve evenly convex with umbo extending slightly posterior of hinge line. Interarea straight orthocline to weakly apsacline, wide and moderately high; delthyrium long and wide. Ornamentation unequally parvicostellate; ribs closely spaced. Few pits developed on ribs. Concentric ornament accentuated at mid-valve and near the anterior margin.

Ventral interior with relatively large teeth supported by recessive dental plates fused onto valve floor to form a callus; callus thickened and continuing anteriorly as small ventral thickening or septum. No other internal features seen.

Remarks. – Only a few pits are visible on the costae of the Alaskan species; possibly most of the pitted ornament is covered by glue. This material, consisting of two ventral valves, is very similar to the species of *Doleroides* found in the Telsitna Formation of the Farewell Terrane (Measures *et al.* 1992; Rohr *et al.* 1992). This is based on the ornament, the rectangular outline, the shape of the ventral interarea and the delthyrial cavity. Measures *et al.* (1992) and Rohr *et al.* (1992) noted that the species closely resembles *D. panna* (Andreeva) from the lower part of the Mangazeya and the Tchertovskaya formations (lower Sandbian) of the Siberian Platform (Andreeva & Nikiforova 1955). They separated the Alaskan species from the Siberian material based on differences in the dorsal cardinalia. As no dorsal valves occur within the current material, open nomenclature for the new species is retained here.

Subfamily Orthostrophinae Schuchert & Cooper, 1931

Genus *Gelidorthis* Havlíček, 1968

Type species. – *Orthis partita* Barrande, 1879; from the Caradoc of Bohemia.

Biogeographic distribution. – Lower Sandbian of Portugal (Botquelen & Mélou 2007); Sandbian–lower Katian of northeastern Spain (Villas 1992), central Spain and possibly also central Europe (Boucot *et al.* 2003).

Gelidorthis perisiberiaensis n. sp.

Plate 8, figures 20–24; Plate 9, figure 1

Derivation of name. – The species name refers to the genus first occurrence close to the Siberian palaeocontinent.

Holotype. – MGUH 29566 (Pl. 8, figs 23, 24).

Paratypes. – MGUH 29564 (Pl. 8, fig. 21), MGUH 29565 (Pl. 8, fig. 22) and MGUH 29567 (Pl. 9, fig. 1).

Material. – One pair of conjoined valves, 46 ventral valves and 29 dorsal valves. All from blocks 4–6. locality A-1230.

Diagnosis. – Shell medium-sized, ventri-biconvex in profile and transverse, subcircular to subtrapezoidal in outline; shell alate in juvenile stages. Anterior commissure unisulcate. Ornamentation parvicostellate. Teeth large, supported by divergent dental plates. Diductor muscle scars bilobed, weakly impressed on callus. Cardinal process thick, bilobate, positioned on notothyrial platform resting on valve floor. Dorsal median septum short. Dental sockets deep; brachiophores low and divergent.

Description. – Shell medium-sized, ventri-biconvex in profile and transverse, subcircular to subtrapezoidal in outline. Most specimens alate, though clearly a result of ontogeny, as growth lines reveal alae at earlier stages. Maximum width at hinge line or just anterior of it. Valve 79–85% as long as wide; anterior commissure unisulcate. Ventral valve evenly convex with small beak; interarea high, wide, curved and apsacline; delthyrium high and narrow. Dorsal valve weakly convex, with deep sulcus; interarea low, flat and anacline; notothyrium open, very wide. Ornamentation parvicostellate with relatively widely spaced, rounded to triangular ribs; up to 3 orders of ribs, numbering 35–42 ribs at the anterior margin; growth lines developed on every quarter of valve length.

Ventral interior with large teeth supported by large, divergent dental plates that converge onto valve floor to form callus. Weakly impressed bilobed, diductor muscle scars on callus. Dorsal interior with thick, possibly bilobate, cardinal process becoming blade-like anteriorly, continuing as short dorsal median septum. Dental sockets deep; brachiophores low and relatively divergent; notothyrial platform situated on valve floor.

Remarks. – *Gelidorthis* may be confused with *Jezercia* Havlíček & Mergl, 1982 but differs in having a much higher, wider, ventral interarea with stout teeth. In addition, exopunctae are developed in *Jezercia*. Villas (1992) commented that the two Bohemian species of *Gelidorthis*, *G. praepartita* Havlíček, 1977 and *G. gemina* Havlíček, 1977; the Spanish *G. meloui* Villas, 1985 and the Welsh *G. cennensis* Lockley & Williams, 1981 have acute cardinal extremities in their young stages of growth but become rectangular to slightly obtuse in adult stages. On the other hand, *G. partita* (Barrande, 1879) often has acute to alate cardinal angles throughout ontogeny. The new species loses its alae during ontogeny and therefore cannot be compared with *G. partita*. Thus, those that have obtuse cardinal extremities are adult specimens (compare Pl. 8, figs 21 and 22). Furthermore, the new species is differentiated from *G. carlsi* Villas, 1985 in having slender teeth positioned on a larger interarea, a smaller dorsal interarea and in possessing a dorsal median septum. *Gelidorthis* usually occurs in Gondwanan siliciclastic deposits (Villas 1992; Botquelen & Mélou 2007). Thus, this is the first report of the genus from lower palaeo latitudes.

Family Cyclocoeliidae Schuchert & Cooper, 1931

Cyclocoeliididae gen. *et* sp. indet.

Plate 9, figures 2–4

Material. – Four ventral valves from blocks 4–6, locality A-1230.

Description. – Ventral valve strongly convex, transversely semi-oval in outline; skewed to the right when viewed perpendicular to exterior, but may be due to deformation. Maximum width at mid-valve length; anterior commissure rectimarginate. Ventral valve unevenly convex, more convex towards anterior margin; umbo positioned posterior to hinge line; interarea curved and strongly apsacline; delthyrium open, high and wide. Ornamentation consisting of a maximum of 15 simple, widely spaced, triangular costae.

Ventral interior with large, wide deltidiodont teeth supported by relatively large dental plates; dental plates converging onto valve floor into a callus extending to about mid-valve as a thickening of the shell. Diductor muscle scars long, narrow and weakly impressed.

Remarks. – The collected specimens are assigned to the Cyclocoeliidae based on its distinctive costate ornamentation and the very narrow hinge line.

Family Cremnorthidae Williams, 1963

Genus *Phragmorthis* Cooper, 1956

Type species. – *Phragmorthis buttsi* Cooper, 1956; from the Darriwilian–lower Sandbian Effna–Rich Valley formations, Virginia, USA.

Biogeographic distribution. – Lower Sandbian of the Midland Valley Terrane (Girvan, Williams 1962); Sandbian–lower Katian of the Chu-Ili Terrane, Kazakhstan (Popov *et al.* 2002; Popov & Cocks 2006) and Virginia and Tennessee, eastern USA (Cooper 1956); Sandbian–Rawtheyan of the Klamath region (Potter 1990a); same author reports the *P.* aff. *buttsi* from the Katian part of the Jones Ridge succession near the Alaska–Yukon border. However, these data are unpublished (Potter & Boucot 1992).

Phragmorthis buttsi Cooper, 1956

Plate 9, figures 5–9

> 1956 *Phragmorthis buttsi* Cooper, new species, p. 510, pl. 146, figs 31–38; pl. 148, figs 5–26; pl. 221, figs 16– 22.
> 1962 *Phragmorthis buttsi* Cooper; Williams, p. 130, pl. XI, figs 40–44, 46–48.
> 1990a *Phragmorthis* aff. *buttsi* morphotype 1 Potter, p. 112, pl. 7, figs 1–29.
> 1990a *Phragmorthis* aff. *buttsi* morphotype 2 Potter, p. 114, pl. 7, figs 31–35.
> 1990a *Phragmorthis* cf. *buttsi* Potter, p. 116, pl. 7, figs 36–56.

Material. – One dorsal valve from block 3 and three pairs of conjoined valves, 36 ventral and 47 dorsal valves from blocks 4, 5 and 5, locality A-1230. Further, one conjoined pair and two dorsal valves from locality 79WG19.

Description. – Shell strongly ventri-biconvex in profile and transverse to subcircular or subquadrate in outline. Maximum width at hinge line; anterior commissure weakly sulcate. Ventral valve unevenly convex, almost triangular to subpyramidal with marked umbo posterior to hinge line. Interarea high, catacline to weakly apsacline; delthyrium high and narrow. Dorsal valve moderately convex, sulcate; dorsal interarea high, wide and strongly anacline; notothyrium narrow. Ornamentation multicostellate with costellae sporadically swollen; concentric growth lines variably developed, increasing in number towards the anterior commissure.

Ventral interior with large deltidiodont teeth supported by dental plates that converge to form spondylium simplex. Dorsal interior lacking cardinal process, or with one very small ridge developed anteriorly on notothyrial platform. Brachiophores long with supporting plates converging to dorsal median septum, more elevated towards the anterior margin. Thin, deep grooves on valve floor.

Remarks. – Apart from its possible occurrence in Jones Ridge, this species is characteristic of exotic terranes such as the Eastern Klamath, Midland Valley- and Chu-Ili terranes.

Family Rhactorthidae Williams, 1963

Genus *Rhactorthis* Williams, 1963

Type species. – *Rhactorthis crassa* Williams, 1963; from the upper Sandbian – lower Katian Gelli-grîn Group, Bala, Wales.

Biogeographic distribution. – The genus is reported from the lower–upper Sandbian of Avalonia (Williams 1963; Parkes 1994); upper Sandbian of Perunica (Havlíček 1977); lower Katian of Baltica (Harper *et al.* 1984) and the Cautleyan of the Midland Valley Terrane (Pomeroy, Wright 1964).

Rhactorthis sp.

Plate 9, figures 10–14

Material. – Nine ventral valves and seven dorsal valves from blocks 4–6, locality A-1230.

Description. – Shell medium-sized, ventri-biconvex, evenly convex profile in both valves and semi-oval outline; cardinal angles obtuse, maximum width probably within one-fifth of valve length from hinge. Umbo posterior to hinge line. Ventral interarea high, curved and apsacline; dorsal interarea narrow, moderately high and anacline; notothyrium open and wide. Ornamentation unequally parvicostellate that consist of thin and closely, but evenly, distributed even-sized primary costae and slightly less pronounced secondary costellae. Ornamentation, with stronger concentric ornamentation near the anterior commissure. Anterior commissure weakly sulcate.

Ventral interior with large, divergent dental plates that converge onto valve floor to form callus situated within a long, deep delthyrium surrounded by the large dental plates. Callus extends to about mid-valve. Callus thickened medianly in the umbonal region, but not continued as median septum. No muscle impressions preserved. Dorsal interior with a thin, low cardinal process, situated on valve floor. Brachiophore bases narrow, but brachiophores more divergent anteriorly; dental sockets small. Low median septum developed anterior of the notothyrial platform extending to mid-valve. Muscle scars not impressed.

Remarks. – Though only few shells occur within the current material, these are quite distinctive and thus ascribed to *Rhactorthis*.

Family Giraldiellidae Williams & Harper, 2000

Genus *Scaphorthis* Cooper, 1956

Type species. – *S. virginiensis* Cooper, 1956; from the lower Sandbian Chatham Hill Formation of Virginia, Eastern USA.

Biogeographic distribution. – Lower Sandbian – upper Katian of the Midland Valley Terrane (Mitchell 1977; Clarkson *et al.* 1992; Candela 2003; Harper 2006; Candela & Harper 2010); Sandbian–lower Katian of British Columbia, Canada (Jin & Norford 1996); lower Katian of New South Wales, Australia (Percival 2009). Potter (1990a) reported the genus from the Katian of Jones Ridge, Yukon Territory and from the Nixon Fork Subterrane (locality 79WG19). However, these data are unpublished and the occurrence from locality 79WG19 cannot be confirmed.

Scaphorthis virginiensis Cooper, 1956

Plate 9, figures 15–22

1956 *Scaphorthis virginiensis* Cooper, new species, p. 505, pl. 55A, figs 1–11.
1962 *Scaphorthis* cf. *virginiensis* Cooper; Williams, p. 129, pl. XI, figs 35–39, 45.

Material. – Three pairs of conjoined valves, 43 ventral and 73 dorsal valves. All from blocks 4–6, locality A-1230.

Description. – Shell ventri-biconvex, subcircular in outline; maximum width at one-third of valve length from posterior margin. Anterior commissure rectimarginate to weakly sulcate. Ventral valve evenly convex; interarea moderately high and curved, strongly apsacline; delthyrium high and relatively wide. Dorsal valve weakly convex, with sulcus at the anterior margin; interarea wide, catacline; notothyrium wide. Ornamentation multicostellate with one to two orders of costellae on left side the primary costae; 39 costellae at the anterior margin. One or two growth lines variably developed near the anterior margin. Impunctate.

Ventral interior with large deltidiodont teeth supported by large dental plates that converge onto valve floor to form callus. No muscle impressions. Dorsal interior with weakly elevated, thin blade-like cardinal process; brachiophores closely spaced, divergent; brachiophore bases converging onto moderately elevated notothyrial platform with wide, relatively large crural plates; crural plates merge anteriorly of the brachiophores continuing into an elevated, relatively thick dorsal median septum. Deep dental sockets posterior of brachiophores.

Remarks. – The current material is assigned to the type species of *Scaphorthis* based on, in particular, the identical multicostellate ornamentation with two to three orders of costellae and one or two very characteristic growth lines just before the anterior commissure. Compared to most other of Cooper's species, the type species has more rounded and coarse costellae. Internally, the notothyrial cavity is deeper and possesses a very subdued blade-like cardinal process; both characters are unlike those of any of the other species referred to this genus. Some of the specimens, described by Potter (1990a) as *Scaphorthis* sp. 1, may, as he also suggested, eventually be assigned to the type species. However, most of his specimens appear to have a different ornament.

Of the other genera within the current material, *S. virginiensis* differs from *Taphrorthis immatura* in outline, being more subcircular, and with growth lines confined to the anterior most part of the shell. Moreover, the cardinal process is less sharp and sessile.

Finally, *Scaphorthis virginiensis* does not possess a ventral median septum.

Superfamily Orthoidea Woodward, 1852
Family Orthidae Woodward, 1852

Genus *Diochthofera* Potter, 1990a

Type species. – *Diochthofera conspicua* Potter, 1990a; from the Klamath Mountains, northern California.

Biogeographic distribution. – Middle Katian–Hirnantian of the Klamath region, Northwest Territories and Jones Ridge, Yukon Territory, both Canada (Potter 1990a).

Diochthofera aff. *conspicua* Potter, 1990a

Plate 9, figures 23–24; Plate 10, figures 1–2

   1990a *Diochthofera conspicua* Potter, p. 60, pl. 1, figs 1–26, 32, 33.

Material. – Three dorsal valves from locality 79WG19.

Description. – Small, convex, sulcate dorsal valve; interarea narrow, flat; notothyrium wide and open. Ornamentation fascicostellate. Interior with low, bulbous cardinal process more elevated anteriorly; large dental sockets. Brachiophores divergent and elevated anteriorly. Adductor field bounded laterally by ridges to form triangular area just anterior of the notothyrial platform. Low dorsal median septum.

Remarks. – This species is probably conspecific with the type species. However, the material is not well preserved, which is probably why Potter (1990b) only listed this locality under the distribution of the genus, and not the type species. This genus is confined to the western margin of present-day North America.

Genus *Taphrorthis* Cooper, 1956

Type species. – *Taphrorthis emarginata* Cooper, 1956, from the lower Sandbian Little Oak Formation, Alabama, USA.

Biogeographic distribution. – Also known from the lower Sandbian of Laurentia (Cooper 1956), the Midland Valley Terrane (Girvan, Reed 1917; Williams 1962) and South China (Chang 1983). In addition the genus has been reported from the Cautleyan of

Pomeroy of the Midland Valley Terrane (Wright 1964; Mitchell 1977) and the middle–upper Katian of the Eastern Klamath Terrane (Potter 1990a). The oldest occurrence is from the lower Darriwilian of Ireland (Williams & Curry 1985), is here also noted, as it is regarded conspecific with the current material.

Taphrorthis immatura Williams & Curry, 1985

Plate 10, figures 3–11

   1977 *Taphrorthis?* sp. Mitchell, p. 30, pl. 30, 31.
   1985 *Taphrorthis immatura* sp. nov. Williams & Curry, p. 228, figs 107–109.

Material. – Five ventral, three dorsal and five fragments from block 3 and one pair of conjoined valves, 126 ventral valves, 70 dorsal valves and three fragments from blocks 4-6 from locality A-1230.

Description. – Shell ventri-biconvex in profile and transverse, subcircular in outline. Maximum width at one-third of valve length from posterior margin; anterior commissure rectimarginate to weakly sulcate. Ventral valve evenly convex; interarea short, curved and weakly apsacline; delthyrium high and relatively wide. Dorsal valve weakly convex, with sulcus deeper at the anterior margin in some specimens; dorsal interarea wide, catacline; notothyrium wide. Ornamentation ramicostellate with a pair of first-order costellae on either side of the primary costae; about 20 primary costae developed in the umbonal region, within one-fourth of valve length from the posterior margin. Growth lines variably developed at the anterior margin.

   Ventral interior with large deltidiodont teeth supported by large dental plates that converge onto valve floor to form callus; muscle scar strong and bilobed, divided by a 'bulge' continuing anteriorly beyond the callus to form a short ventral, median septum. The median septum extends to mid-valve length. Dorsal interior with strong, blade-like cardinal process more elevated anteriorly on the notothyrial platform. Brachiophores relatively narrow, but divergent, converging into an elevated notothyrial platform with wide, relatively large crural plates. Deep dental sockets visible, posterior to brachiophores. Notothyrial platform supported anteriorly by a relatively large, wide dorsal median septum extending anteriorly until mid-valve valve.

Remarks. – The material appears punctate, however this could be due to the glue used to assemble shell fragments. Apart from the general outline, this species is referred to *Taphrorthis* on the basis of the raised notothyrial platform, the well-developed ventral median ridge in most specimens and the characteristic ramicostellate ornamentation.

Previous records of the genus are almost exclusively characterized by insufficient material, with only a few shells available for study (Williams 1962; Wright 1964; Mitchell 1977; Potter 1990a). Thus, the current study probably presents the most abundant material of the genus and thus intraspecific variability can be better assessed.

Some characters have consistently been discussed by the above-mentioned studies. These are the absence or presence of a dorsal sulcus, filate ornament and the development of a ventral median adductor muscle ridge. Various combinations of these have formed the basis for definition of the different species within the genus.

This larger collection of specimens show that the dorsal sulcus is only developed in the most posterior part of the dorsal valve, giving juvenile specimens a sulcate appearance, whereas in adult specimens the anterior commissure is rectimarginate. A shallow sulcus in juvenile specimens was also observed in *T. pecularis* Cooper, 1956; *T.* aff. *pecularis* Williams, 1962; *Taphrorthis?* sp. Wright, 1964; *T. immatura* Williams & Curry, 1985; *T. emarginata* Cooper, 1956; and *T. bellatrix* Williams, 1962, but with the difference that the anterior commissure instead becomes uniplicate in adult specimens of the last two species. Another feature that seems to be related to growth is the development of a raised notothyrial platform in adult specimens, compared to a sessile platform in juveniles (see Pl. 10, figs 1–4).

The characteristic ventral median ridge is very clearly developed in most specimens, although almost lacking in a few specimens and thus cannot with absolute certainty be species diagnostic. Another character that seems to be diagnostic for this material is the development of concentric ornament, or growth rings, in the anterior half of the shell. This trait is consistent in several species within the genus. But the current material does not contain any specimens that have a filate ornament.

Thus, summarized, the current material is not conspecific with the type species *T. emarginata* and *T. bellatrix*, which both possess filae and have an uniplicate anterior commissure. *T. pecularis* and *T.?* sp. from the Portrane Limestone (Wright 1964), also possesses filae, and thus are not considered conspecific either. Of the species that are not filate, that is *T. aspera* Williams, 1962; *Taphrorthis?* sp. Mitchell, 1977; *Taphrorthis?* sp. Potter, 1990a; *T. immatura* Williams & Curry, 1985; the current material is almost identical to *T. immatura* from the lower Darriwilian Tourmakeady Limestone of Ireland. We have further tentatively synomnized our material with the upper

Katian *Taphrorthis?* sp. from the Killey Bridge Formation of Pomeroy (Mitchell 1977), based on the appearance of the dorsal cardinalia in the material from Northern Ireland (Pl. 2, figs 30, 31). This, however, is speculative; it would extend the species a range from the lower Darriwilian to the upper Katian.

Family Plaesiomyidae Schuchert, 1913

Subfamily Plaesiomyinae Schuchert, 1913

Genus *Austinella* Foerste, 1909

Type species. – *Orthis kankakensis* McChesney, 1861; from the middle–upper Katian Maquoketa Formation of Iowa, USA.

Biogeographic distribution. – Lower–middle Sandbian of the Chu-Ili Terrane (Popov *et al.* 2002) and the York Terrane, Seward Peninsula, western Alaska (Potter 1984; Potter *et al.* 1988); upper Sandbian – lower Katian of the Chinghiz Terrane (Klenina *et al.* 1984); lower Katian – middle Katian (Pusgillian) of Altai (Severgina 1978; Sennikov *et al.* 2008); middle–upper Katian (Pusgillian–Rawtheyan) of Laurentia (Howe 1966, 1988) and the Eastern Klamath Terrane (Potter 1990a).

Austinella? sp.

Plate 10, figures 12–15

Material. – One ventral valve and two dorsal valves from blocks 4–6, locality Λ-1230.

Description. – Shell ventri-biconvex with a sulcate anterior commissure. Ventral interarea high and orthocline; dorsal interarea plane, short and weakly anacline. Delthyrium and notothyrium open, the latter very wide. Ornamentation consists of ramicostellae, with sharp and thin costellae that are relatively, widely spaced. Second order costellae developed in the most anterior part of the valve only. Sporadic aditicules developed.

Ventral interior with wide dental plates converging onto valve floor to form a callus that possess a thick, raised, quadrate area, medially. Teeth not preserved. Dorsal interior with blade-like cardinal process, more elevated towards hinge line. Brachiophores high and wide; brachiophore plates slightly curved anteriorly to form semi-oval notothyrial platform that is situated almost on valve floor.

Remarks. – Although the material is sparse, assignment is made with some confidence based ornamentation and cardinalia in both the ventral and the dorsal valves, respectively.

Genus *Dinorthis* Hall & Clarke, 1892

Type species. – *Orthis pectinella* Emmons, 1842 from the upper Sandbian – lower Katian Trenton Limestone of New York, USA.

Biogeographic distribution. – Lower Sandbian of Bolivia (Suárez-Soruco 1992) and Wales (Williams 1963); Sandbian of New South Wales (Percival 1991); lower Sandbian – lower Katian of the Midland Valley Terrane (Williams 1962; Candela 2003); upper Sandbian – lower Katian of Tasmania (Laurie 1991; Nikitin *et al.* 2003); lower Katian of British Columbia, Canada (Jin & Norford 1996); eastern USA (Patzkowsky & Holland 1997), northern Kyrgyzstan, Selety River Basin and Chu-Ili Terrane, Kazakhstan (Popov *et al.* 2000; Nikitin *et al.* 2003); lower–middle Katian of southern Manitoba (Jin & Zhan 2000) and the middle–upper Katian of eastern. North Greenland (Rasmussen & Harper 2010) and South China (Popov *et al.* 2000; Jin *et al.* 2007).

Dinorthis sp.

Plate 10, figures 16–21

Material. – Eight ventral, 24 dorsal and two unidentifiable fragments, all from blocks 4–6, locality A-1230.

Description. – Shell large, ventri-biconvex in profile, transversely semi-oval in outline. Cardinal angles obtuse; maximum width at one-third to half of valve length from posterior margin; anterior commissure rectimarginate to weakly sulcate. Ventral valve evenly convex; interarea moderately high, narrow and curved, weakly apsacline; delthyrium high and relatively wide. Dorsal valve weakly convex, with sulcus deeper at anterior margin; dorsal interarea high, strongly anacline; notothyrium open and very wide. Ornamentation consists of simple costae to weak parvicostellae (ribs bifurcate in the umbonal region in some specimens). Costae are high, rounded, numbering 32 at the anterior margin, appear filate in depressions. One growth line developed near the anterior margin in most specimens.

Ventral interior with large deltidiodont teeth supported by large dental plates that converge onto valve floor into a weak callus; diductor scars strong, bilobed and divided medianly by a small rounded elevation. Externally from the diductor muscle scars, small muscle bounding ridges are developed, dividing the muscle scars into four quarters. Dorsal interior with small blade-like cardinal process; brachiophores very wide and divergent,

with bases converging into notothyrial platform; wide, relatively large crural plates merging anteriorly of the brachiophores and continuing into a short, low dorsal median septum, extending anteriorly to mid-valve. Deep dental sockets posterior to brachiophores.

Remarks. – The present specimens are separated from other orthide genera by the bilobed ventral muscle scars, as well as the simple, costate ornamentation. The costae appear higher and more rounded than the specimens illustrated in the Treatise (Williams & Harper 2000) and as the current specimens are smaller than those typical for the genus, this may eventually be grounds for erecting a new species.

Family Hesperorthidae Schuchert & Cooper, 1931

Genus *Ptychopleurella* Schuchert & Cooper, 1931

Type species. – *Orthis bouchardi* Davidson, 1847 from the Much Wenlock Limestone Formation, Bethnal Ridge, Shropshire, UK.

Biogeographic distribution. – Reported from the lower Sandbian – upper Katian of the Midland Valley Terrane (Williams 1962; Mitchell 1977; Harper 1986, 2000, 2006; Candela 2003); Sandbian of Virginia, USA (Cooper 1956); Sandbian–lower Katian of Australia (Laurie 1991; Percival 1991, 2009) and Myanmar (Cocks & Zhan 1998); lower Katian of Kazakhstan, including the Chu-Ili Terrane (Popov & Cocks 2006); upper Sandbian–Cautleyan of northeastern Spain (Villas 1985); upper Cautleyan–Rawtheyan of Sweden (Wright 1982); Rawtheyan of Belgium (Sheehan 1987) and Taimyr, North Russia (Cocks & Modzalevskaya 1997); Katian of Northwestern France (Mélou 1990) and Sardinia, Italy (Leone *et al.* 1991). Finally, Ross & Dutro (1966) reported the genus from Jones Ridge, east-central Alaska and Potter & Boucot (1992) reported the current material from the White Mountain area as well as Yukon Territory. As already mentioned these reports were not particularly well age constrained.

Ptychopleurella uniplicata Cooper, 1956

Plate 10, figures 22–24; Plate 11, figures 1–3

1956 *Ptychopleurella uniplicata* Cooper, new species, p. 391, pl. 49F, 41–44.

1966 *Ptychopleurella* cf. *lapworthi* Ross & Dutro, p. 5, pl. 1, figs 2, 4, 6, 8.

1990a *Ptychopleurella* aff. *uniplicata* Potter, p. 94, pl. 4, figs 57–66; pl. 5, figs 1–30.

Material. – Two dorsal valves from block 3 and 17 pairs of conjoined valves, 114 ventral and 132 dorsal valves from blocks 4–6, locality A-1230. Two pairs of conjoined valves, four ventral, five dorsal and to indeterminable fragments from locality 79WG19. One dorsal valve from locality 79WG126.

Description. – Shell small to medium-sized; ventri-biconvex profile; transversely semi-rectangular outline. Cardinal angles obtuse; maximum width one-third to half of valve length from posterior margin; anterior commissure rectimarginate to weakly unisulcate. Ventral valve strongly convex, subpyramidal; specimens with weak sulcus have a weak bilobed outline. Interarea high and cataline, straight; delthyrium high and very narrow. Dorsal valve weakly convex, with deeper sulcus at the anterior margin; interarea relatively high, wide and strongly anacline; notothyrium open, large and wide. Ornamentation consists of 11–13 coarse, simple costae that are interrupted by a strong lamellose concentric ornamentation; ornament of 5 costae in the ventral sulcus.

Ventral interior with large teeth supported by relatively large dental plates that fuse with valve floor to form a strong callus containing bilobed diductor muscle scar impressions. Anterior of the callus, the median is elevated as a result of a small depression in the ventral valve in some specimens. Dorsal interior with small blade-like cardinal process. Dental sockets large and deep; brachiophores long, stout and moderately divergent. Notothyrial platform elevated; thick dorsal median septum extends anteriorly to commissure.

Remarks. – The current specimens closely resemble those of *P.* aff. *uniplicata* from the Klamath Mountains described by Potter (1990a), but also of *P.* cf. *lapworthi* (Davidson 1883) from Jones Ridge, east-central Alaska (Ross & Dutro 1966). The current study therefore also suggests assigning the Jones Ridge material of *P.* cf. *lapworthi* to *P. uniplicata* on the basis of fewer costae and a straight ventral interarea in *P. uniplicata* as compared to more frequent costae and the curved ventral interarea in *P. lapworthi* (Harper 1986). Due to the small number of specimens in the upper Katian material (79WG19) we also tentatively assign these to *P. uniplicata*. The Jones Ridge material is probably conspecific with the White Mountain material.

Family Glyptorthidae Schuchert & Cooper, 1931

Genus *Glyptorthis* Foerste, 1914

Type species. – *Orthis insculpta*, Hall 1847 from the Katian Richmond Group of Ohio, USA.

Biogeographic distribution. – Reported from the lower Sandbian of South China (Rong *et al.* 1999); Sandbian–lower Katian of Gorny Altai, Russia (Kulkov & Severgina 1989), the Chu-Ili Range, Kazakhstan (Popov *et al.* 2002), Pomeroy (Mitchell 1977; Candela 2003), Myanmar (Zhan & Cocks 1998) and South China (Rong *et al.* 1999; Chen *et al.* 2000); upper Sandbian – lower Katian of the Oslo Region (Hansen 2008); lower Katian of British Columbia, Canada (Jin & Norford 1996), Kilbucho and Wallace's Cast, Scotland (Clarkson *et al.* 1992; Candela & Harper 2010), Shaanxi, North China (Rong *et al.* 1999) and Siberia (Kanygin *et al.* 2006); lower–middle Katian of Québec (Sheehan & Lespérance 1979) and Sweden (Cocks 2005); upper Cautleyan–Rawtheyan of Iowa, Ohio, Oklahoma, Western Texas, Tennessee (Howe 1988); middle Katian–Hirnantian of Girvan, Scotland (Cocks 1988; Harper 2006); Rawtheyan of Belgium (Sheehan 1987), Wales, Southern Poland and Sardinia, Italy (Villas *et al.* 2002). Additionally, the genus also occurs in the Hirnantian of the Oslo Region (Cocks 1988).

Glyptorthis sp.

Plate 11, figures 4–9

Material. – 36 ventral and 15 dorsal valves from blocks 4, 5 and 6, locality A-1230 and one ventral and four dorsal valves from locality 79WG126.

Description. – Shell medium to large, subequally biconvex with slightly ventri-biconvex profile and subquadrate outline. Cardinal angles obtuse, maximum width within one-quarter of valve length from posterior margin; anterior commissure rectimarginate. Ventral valve gently convex. Interarea high, wide and cataline; delthyrium high and slightly narrow. Dorsal valve weakly convex, with variably developed sulcus. Interarea relatively low, linear and strongly anacline; notothyrium open, large and wide. Ornamentation unequally ramicostellate, numbering 15–22 at the anterior margin. Strong concentric ornamentation, towards the anterior margin becomes lamellose.

Ventral interior with large teeth supported by relatively large dental plates that converge onto valve floor to form callus on which elongate, thin diductor muscle scars are deeply impressed. Callus elevated at its anteriormost end on some specimens. Dorsal interior with small, blade-like cardinal process flanked by elevated brachiophores; dental sockets deep and wide. Brachiophores long and wide, bounding triangular notothyrial platform. Thin median septum, more elevated at its anterior end, developed anterior to the cardinal process.

Remarks. – These specimens closely resemble *Glyptorthis* sp. 1 of Potter (1990a), except for the catacline ventral interarea. The specimens could easily be confused with *Ptychopleurella uniplicata* but they differ in size and have very different interiors.

Family Hesperorthidae Schuchert & Cooper, 1931

Genus *Hesperorthis* Schuchert & Cooper, 1931

Type species. – *Hesperorthis tricenaria* Conrad, 1843; from the lower Katian Trenton Limestone of Wisconsin, USA.

Biogeographic distribution. – This genus has a near cosmopolitan distribution. Reported from the lower Sandbian of Newfoundland (McKerrow & Cocks 1981) and Perunica (Havlíček 1977); lower–upper Sandbian of the Northern Precordillera (Benedetto 2002); lower Sandbian – lower Katian of New South Wales and Tasmania (Laurie 1991; Percival 1991, 2009); lower Sandbian – middle Katian of Altai (Kulkov & Severgina 1989; Sennikov 2008); lower Sandbian – upper Katian of Avalonia (Carlisle 1979; Hiller 1980; Parkes 1992; Harper & Brenchley 1993), Eastern Klamath Terrane (Potter 1990a), the Midland Valley Terrane (Williams 1962; Clarkson *et al.* 1992; Candela 2003; Harper 2006; Candela & Harper 2010) and Siberia including Taimyr (Nikiforova & Andreeva 1961; Cocks & Modzalevskaya 1997; Kanygin *et al.* 2006); lower Sandbian–Hirnantian of Baltica (Sheehan 1979; Cocks 1988; Paškevičius 1994; Neuman *et al.* 1997) and Laurentia (Cooper 1956; Howe 1966; Macomber 1970; Alberstadt 1973; Jin & Zhan 2008). In addition the genus is reported from the Hirnantian of Kolyma (Cocks 1988).

Hesperorthis sp.

Plate 11, figures 10–12

Material. – One ventral valve from locality 79WG126.

Description. – Small, strongly convex profile with a rectangular, slightly bilobed outline. Maximum width approximately at mid valve. Ornament of estimated18 simple, slightly angular, coarse costae, two of which are developed in the small depresseion medially. A pitted ornament appears to be distributed randomly both on and between the costae. Interarea very high and wide, delthyrium narrow. Teeth small.

Ventral interior with deep delthyrial cavity bounded by recessive dental plates that join on valve floor to form weakly elevated callus. Muscle field cordate.

Remarks. – Differentiated from *Hesperorthis* sp. 2 on the basis of the very high ventral interarea. Chiang (1972) surveyed most of the species within the genus. However, none of those depicted in that study appear to have the same very high ventral interarea that is so characteristic for the current Alaskan species. As no dorsal valves have been found, this species has been left under open nomenclature.

Hesperorthis sp. 2

Plate 11, figures 13–17

Material. – Nine ventral and ten dorsal valves. All from blocks 4–6, locality A-1230.

Description. – Shell medium-sized, ventri-biconvex, weakly unisulcate and subcircular in outline. Ornamentation consisting of simple, rounded costae numbering 20–25 at anterior margin; valves are punctate. Some specimens with weak capillae. Ventral interarea high, narrow, straight and orthocline; open long and narrow delthyrium. Dorsal interarea low, flat, and weakly anacline; notothyrium open, wide.

Ventral interior with large teeth supported by recessive, narrow dental plates that converge onto valve floor to form low callus. Dorsal interior with a small, thin, blade-like cardinal process and wide brachiophores that are supported by plates converging to form a low notothyrial platform, that continues anteriorly as a thickening of the shell. Dental sockets small and thin.

Remarks. – The present specimens could be confused with *Oanduporella kuskokwimensis*. However, the Alaskan specimens differ in having a higher and orthocline ventral interarea and different cardinalia. Furthermore, the punctae are not as large and numerous as in *Oanduporella kuskokwimensis*.

Family Orthidae Woodward, 1852

Genus *Duolobella* n. gen.

Type species. – *D. sandiae* locality A-1230 (upper Darriwilian — lower Sandbian) of the White Mountain area, west-central Alaska.

Derivation of name. – The name of the new genus refers to its characteristic bilobed outline.

Diagnosis. – Shell small, biconvex, with transversely lobate outline. Anterior commissure unisulcate. Cardinal angles obtuse. Interarea moderately high. Delthyrium high. Teeth flat and wide, supported by divergent dental plates. Dorsal interarea low, notothyrium open. Ornament costate. Brachiophores widely divergent. Notothyrial platform absent. Cardinal process small and thick.

Description. – Shell small, biconvex with transversely lobate outline, about 50% as long as wide. Cardinal angles obtuse; maximum width approximately at midvalve; anterior commissure strongly unisulcate. Ventral valve convex. Interarea moderately high, wide, and catacline to weakly apsacline; delthyrium high and moderately wide. Dorsal valve strongly convex. Interarea low, flat and catacline; notothyrium open, wide. Ornamentation of 25 simple, flat and rounded costae to weakly parvicostellate along the anterior margin; ribs widely spaced. Strong depression developed medianly, both on the ventral fold and the dorsal sulcus. Up to two growth lines developed in the anterior half of the valve.

Ventral interior with large, flat and wide teeth supported by small, divergent dental plates, converging onto valve floor to form weakly elevated callus. Dorsal interior with deep sockets and widely divergent brachiophores. Notothyrial platform absent. Cardinal process small, represented by thick bulge.

Duolobella sandiae n. gen. et n. sp.

Plate 11, figures 18–24; Plate 12, 1–2

Derivation of name. – Named after Sandie Mosdal.

Holotype. – MGUH 29607 (Pl. 11, figs 18–20, 23, 24).

Paratypes. – MGUH 29608 (Pl. 11, figs 21, 22), MGUH 29609 (Pl. 12, fig. 1) and MGUH 29610 (Pl. 12, fig. 2).

Material. – 16 ventral and four dorsal valves. All from locality A-1230, blocks 4–6.

Diagnosis. – As for genus.

Description. – As for genus.

Remarks. – This genus is placed within the Orthida, based on the blade-like cardinal process and the lack of fulcral plates (that are often present in plectorthoids). Although the cardinalia resembles that of *Dicoelosia*, the valves are impunctate, and are thus referred to the Orthidina. However, a pitted ornament is often lost during the process of silification (Jisuo Jin, personal communication, 2011). Therefore, this new genus may eventually be transferred to the Dicoelosiidae. Nevertheless, outline and profile clearly separate the present genus from *Dicoelosia*.

Duolobella is a small orthoid within the present fauna that seems to have been particularly well suited for this deep-water setting.

Genus *Palaeowingella* n. gen.

Type species. – *Palaeowingella farewellensis*; from upper Darriwilian – lower Sandbian of the White Mountain area, west-central Alaska.

Derivation of name. – Due to the alate, wing-like cardinal angles.

Diagnosis. – Shell small, ventri-biconvex with transversely lobate to triangular outline. Cardinal angles alate. Ventral interarea moderately high, dorsal interarea low. Delthyrium long, covered by pseudodeltidium. Teeth large. Notothyrium open and wide. Ornamentation consists of simple, widely spaced costae. Dental sockets deep; brachiophores widely divergent. Cardinal process weak, blade-like.

Description. – Shell small, ventri-biconvex with transversely lobate to triangular outline. Cardinal angles alate; maximum width at hinge line. Valves about 60% as long as wide; anterior commissure unisulcate. Ventral valve convex with small umbo. Interarea moderately high, wide, curved and apsacline; delthyrium long and wide, covered by pseudodeltidium. Dorsal valve weakly convex, with deep sulcus. Interarea low, flat and weakly anacline; notothyrium open and wide. Ornamentation consists of simple, widely spaced, costae numbering 22 at anterior commissure. One growth line developed at the anterior commissure.

Ventral interior with large teeth that projects from pseudospondylium. Not supported by median septum. Dorsal interior with wide dental sockets and widely divergent brachiophores that converge into a cardinal process on an weakly elevated notothyrial platform. Cardinal process weak, bladelike, continuing anteriorly as a thin, weak dorsal median septum that becomes more elevated towards the anterior margin.

Palaeowingella farewellensis n. gen. et n. sp.

Plate 12, figures 3–8

Derivation of name. – Named after its occurrence within the Farewell Terrane, west-central Alaska.

Holotype. – MGUH 29611 (Pl. 12, figs 3, 5–7).

Paratype. – MGUH 29612 (Pl. 12, figs 4, 8).

Material. – Four ventral valves and one dorsal valve from blocks 4–6, locality A-1230.

Description. – As for genus.

Remarks. – Although very few specimens were collected, the distinct outline and profile warrant new generic status. However, the phylogenetic placement of this new genus is more difficult. The delthyrial cavity resembles that seen in the Orthidina, as does the blade-like cardinal process and divergent brachiophores. The new taxon is not referred to the Skenidioidea, although the internal characters could suggest an position close to that superfamily. However, we have not placed this new genus in the Skenidioidea due to the lack of apical support for the spondylium and the absence of a true septalium in the dorsal valve.

Family Unknown

Orthoidea gen. *et* sp. indet. 1

Plate 12, figures 9–11

Material. – One ventral valve from locality A-1230, blocks 4–6.

Description. – Shell small and strongly convex with transversely lobate to triangular outline, about 55% as long as wide. Cardinal angles alate; maximum width at hinge line; anterior commissure uniplicate. Ventral valve triangular and lobate. Deep sulcus developed medianly. Interarea moderately developed, curved and catacline; delthyrium high and wide. Ornamentation of simple, accentuated costae to weakly parvicostellate triangular and widely spaced ribs, numbering about 22–24 costellae at anterior margin.

Ventral interior apparently lacking teeth, probably lost due to breakage. Small, divergent dental plates converge onto valve floor to create a callus.

Remarks. – This specimen may represent a new taxon. However, more material is needed to erect a new genus with confidence. It is placed within the Orthoidea due to the lack of a pitted ornament.

Orthidina gen. *et* sp. indet. 1

Plate 12, figures 12–13

Material. – Four dorsal and two indeterminate fragments. All from locality A-1230, blocks 4, 5 and 6.

Description. – Dorsal valve medium-sized, convex and subcircular. Cardinal angles obtuse; maximum width within one-third of the anterior margin; anterior commissure weakly sulcate; sulcus deeper at anterior margin. Dorsal interarea high, wide and anacline; notothyrium open and relatively wide. Ornamentation multicostellate; strongly lamellose on the anterior half of the valve.

Dorsal interior with long and narrow brachiophores supported by brachiophore plates that converge onto valve floor; dental sockets deep. Cardinal process continuing anteriorly as short thickening of valve, but without a median septum.

Remarks. – This may not be an orthoid. However, as the cardinalia is missing it is provisionally positioned here.

Orthidina gen. *et* sp. indet. 2

Plate 12, figures 14–17

Material. – One ventral and two dorsal valve from blocks 4, 5 and 6, locality A-1230.

Description. – Medium-sized shell with plano-convex profile and subquadrate outline. Cardinal angles obtuse, maximum width approximately at mid-valve. Anterior commissure rectimarginate to weakly unisulcate. Ventral valve evenly convex, highest point at umbo. Interarea moderately high, narrow and catacline; delthyrium long and narrow. Dorsal valve flat, with a weak sulcus. Dorsal interarea low, flat and anacline; notothyrium open, wide. Ornamentation unequally ramicostellate, numbering 20-25 at anterior margin, with closely spaced, weak, concentric ornamentation. One growth line near anterior margin.

Ventral interior with large deltidiodont teeth supported by small dental plates. Dorsal interior with thick, rod-like cardinal process situated on valve floor bounded by low, widely divergent brachiophores.

Remarks. – This material resembles *Eridorthis* in ornamentation, but lacks the ventral sulcus or dorsal fold characteristic of that genus. The characteristic ornament also may resemble that characteristic of *Glyptorthis*. However, the dorsal cardinalia is clearly different from that genus. It is possible that the ventral and dorsal valves may not be conspecific; however they resemble one another in both ornamentation and outline.

Orthidina gen. *et* sp. indet. 3

Pl. 12, figures 18–19

Material. – Three dorsal valves from blocks 4–6, locality A-1230.

Description. – Dorsal valve small, convex with transversely semi-oval outline. Cardinal angles at right angle;

maximum width at hinge line; anterior commissure sulcate with sulcus deeper at anterior margin. Interarea high and anacline; notothyrium open and wide. Ornamentation unequally parvicostellate, with one concentric growth line near the anterior margin.

Dorsal interior with thin, blade-like cardinal process situated between narrow brachiophores. Brachiophore plates converge onto elevated notothyrial platform. making the cardinalia of this species characteristic. Dental sockets deep, wide and triangular. Elevated dorsal median septum developed anterior to the notothyrial platform and extending up to anterior margin. Muscle scars not impressed.

Remarks. – The cardinalia of this species is very characteristic.

Orthidina gen. *et* sp. indet. 4

Plate 12, figures 20–22

Material. – One ventral valve from locality A-1230, blocks 4–6.

Description. – Ventral valve small, convex and triangular with probably rectangular outline. Cardinal angles not preserved; umbo at hinge line. Interarea orthocline, wide and high; delthyrium long and narrow. Ornamentation parvicostellate with first order costae and second order costellae arising near the umbonal region. Ventral interior with large teeth supported by recessive dental plates that converge onto valve floor to form callus.

Orthidina gen. *et* sp. indet. 5

Plate 12, figures 23–24

Material. – One ventral valve from blocks 4–6, locality A-1230.

Description. – Ventral valve gently convex; umbo overhanging hinge line. Interarea low, very wide, weakly curved, orthocline to procline; delthyrium open, narrow. Ornamentation consists of simple, even sized costae that are relatively widely spaced numbering 18 in the umbonal region.

Ventral interior with relatively divergent dental plates that converge onto valve floor to form a rounded delthyrial cavity with callus, slightly thickened medianly.

Remarks. – The material is similar to genera of the family Orthidae Woodward, 1852 primarily based on the simple ornamentation and the long, low interarea. The ornamentation, the narrow delthyrium and

arrangement of the dental plates, resemble those seen in *Paralenorthis*. However, this single ventral fragment is too poorly preserved to warrant a confident generic assignment.

Orthidina gen. *et* sp. indet. 6

Plate 13, figures 1–5

Material. – Two ventral valves and two dorsal valves from blocks 4–6, locality A-1230.

Description. – Shell ventri-biconvex, with subcircular outline. Ventral interarea high, orthocline to weakly apsacline; delthyrium open, long. Dorsal interarea high, narrow; notothyrium large, very wide. Ornamentation unequally parvicostellate consisting of simple, thin, sharp costae with intercepting lower second order costellae.

Ventral interior with large teeth, supported by wide dental plates that converge onto valve floor to form callus, slightly elevated medianly. Diductor muscle impressions tear-shaped, narrow in apex, widest towards anterior rim of callus. Dorsal interior with long, thin, blade-like, cardinal process; brachiophores are widely divergent, extending anteriorly. Brachiophore plates fused to form a wide notothyrial platform, anterior of their maximum lateral extension, continued as low, very thick dorsal median septum or thickening.

Remarks. – SEM images have revealed borings in the ventral valve exterior (Pl. 13, fig. 15). This suggests that the shells may have been subject to predation.

Suborder Dalmanellidina Moore, 1952
Superfamily Dalmanelloidea Schuchert, 1913
Family Dalmanellidae Schuchert, 1913
Subfamily Dalmanellinae Schuchert, 1913

Genus *Paucicrura* Cooper, 1956

Type species. – *Orthis rogata* Sardeson, 1892; from the lower Katian of Minnesota, USA.

Biogeographic distribution. – Lower Sandbian of Altai (Sennikov *et al.* 2008); lower Sandbian – upper Katian of the Midland Valley Terrane (Williams 1962; Candela 2003, 2006a; Candela & Harper 2010) and the USA (Sardeson 1892; Cooper 1956; Howe 1988); upper Sandbian–Hirnantian of Baltica (Cocks 1988; Neuman *et al.* 1997; Hansen 2008) and the lower Katian of British Columbia, Canada (Jin & Norford 1996).

***Paucicrura*? sp.**

Plate 13, figures 6–9

Material. – One ventral valve, four? dorsal valves and two fragments from blocks 4, 5 and 6, locality A-1230.

Description. – Shell ventri-biconvex in profile, and probably suboval in outline. Maximum width at hinge line, or within one-third of shell length from posterior margin; anterior commissure weakly unisulcate. Ventral interarea moderately high, catacline; delthyrium high and wide. Dorsal valve with low interarea, probably anacline; notothyrium wide. Ornamentation parvicostellate.

Ventral interior with large teeth. Dorsal interior with deep dental sockets; brachiophores high, sharp; cardinal process small, developed as thick bulge posteriorly. Dorsal median septum lacking. Ring of muscle bounding ridges extending laterally of brachiophores and encircling about half of valve area.

Remarks. – No other internal features are preserved in ventral valve.

Family Dicoelosiidae Cloud, 1948

Genus *Dicoelosia* King, 1850

Type species. – *Anomia biloba* Linnaeus, 1758;. According to Cocks (2008a) a later selected lectotype is of probable Wenlock age from either Gotland, Sweden, or Shropshire, UK.

Biogeographic distribution. – Reported from the upper Sandbian to middle Katian of Pomeroy (Candela 2003; Mitchell 1977) middle Katian of Estonia; Pusgillian–Cautleyan of Girvan, Scotland (Harper 2006); Cautleyan of England (Jin & Copper 1999) and Sweden (Sheehan 1979; Jin & Copper 1999); middle–upper Katian of the Klamath Mountains, northern California (Potter 1990b); upper Cautleyan–Rawtheyan of Québec, Canada (Jin & Copper 1999); Rawtheyan of Belgium (Sheehan 1987) and Taimyr, northern Russia (Cocks & Modzalevskaya 1997); Hirnantian of England and Oklamoma, USA (Cocks 1988). Also reported from Jones Ridge, east-central Alaska by Ross & Dutro (1966).

Dicoelosia jonesridgensis Ross & Dutro 1966

Plate 13, figures 10–12

> 1964   *Dicoelosia* sp. Dutro & Ross, p. 53.
> 1966   *Dicoelosia jonesridgensis* n. sp. Ross & Dutro, p. 6, pl. 3, figs 1–5.

> 1990b   *Dicoelosia jonesridgensis*? Potter, p. 40, pl. 5, figs 48–60; pl. 6, figs 1–21.

Material. – Five pairs of conjoined valves from locality 79WG19.

Description. – Shell small, concavo-convex in profile and bilobate in outline, about 90% as long as wide. Cardinal angles obtuse; maximum width near anterior margin; anterior commissure strongly uniplicate. Ventral valve evenly convex. Interarea relatively high, short and strongly apsacline; delthyrium high and relatively narrow. Dorsal valve concave. Notothyrium open and relatively wide. Ornamentation weakly costate. Internal features not preserved except for a dorsal median septum.

Remarks. – Internally *D. jonesridgensis* is one of only two species within *Dicoelosia* that posseses transverse dorsal ridges (Chen *et al.* 2008). However, as no interior characters are obvious within the current material, this character cannot be used in the present context. The specimens are instead assigned to *D. jonesridgensis* on the basis of the very narrow lobes compared with other known species of this genus.

Genus *Epitomyonia* Wright, 1968

Type species. – *E. glypha* Wright, 1968 from the upper Katian Boda Limestone, Siljan, southern Sweden.

Biogeographic distribution. – Upper Sandbian – lower Katian of Shaanxi, northwest China (Zhan & Jin 2005); lower Katian of Kazakhstan (Popov & Cocks 2006); middle–upper Katian of the Czech Republic (Boucot *et al.* 2003), Klamath Mountains (Potter 1990b) and Spain (Villas 1985); upper Cautleyan–Rawtheyan of Sweden (Sheehan 1979; Cocks 2005); Rawtheyan of Altai (Kulkov & Severgina 1989); Sardinia, Italy (Villas *et al.* 2002), South China (Zhan & Cocks 1998) and Taimyr, North Russia (Cocks & Modzalevskaya 1997); Hirnantian of Norway (Cocks 1988), Missouri, USA and Wales (Zhang & Boucot 1988).

Epitomyonia relicina Potter, 1990b

Plate 13, figures 13–15

> 1990b   *Epitomyonia relicina* n. sp. Potter, p. 46, pl. 6, figs 62–68, pl. 7, figs 1–31.

Material. – One pair of conjoined valves, one ventral and one dorsal fragment. All from locality 79WG19.

Description. – Shell small, ventri-biconvex in profile and longer than wide. Dorsal valve strongly resupinate with bilobate outline. Cardinal angles obtuse; maximum width almost at anterior margin; anterior commissure strongly plicate. Ventral valve evenly convex with small depression medianly in the umbonal region posterior to lobation. Interarea high and strongly apsacline. Delthyrium long, triangular and relatively narrow. Dorsal valve flat to weakly concave at umbonal zone, but becoming strongly concave towards anterior commissure; sulcus deeper at the anterior margin. Interarea low, flat and weakly anacline; notothyrium open and relatively wide. Ornamentation parvicostellate.

Internally, a dorsal median septum and a thick, rod-like cardinal process, are only preserved.

Remarks. – The present specimens resemble some more wide-lobed species of *Dicoelosia* in outline. Furthermore, the long, curved ventral interarea also exists in both *Dicoelosia* and *Epitomyonia*. However, lobes in *Dicoelosia* never are as wide as in *Epitomyonia*. The dorsal interior of the current specimens of *Epitomyonia* also indicate a larger dorsal median septum than what is to be observed in *Dicoelosia*.

Superfamily Enteletoidea Waagen, 1884

Family Draboviidae Havlíček, 1950

Subfamily Draboviinae Havlíček, 1950

Genus *Diorthelasma* Cooper, 1956

Type species. – *Diorthelasma parvum* Cooper, 1956; from the upper Darriwilian Pratt Ferry Formation, Alabama, eastern USA.

Biogeographic distribution. – The genus is known from the upper Darriwilian of Alabama (Cooper 1956), the lower Sandbian – middle Katian of the Midland Valley Terrane (Williams 1962; Mitchell 1977) and the lower Silurian of Oslo (Baarli 1988).

Diorthelasma parvum Cooper, 1956

Plate 13, figures 16–21

1956 *Diorthelasma parvum* Cooper, p. 998, pl. 146, figs 5–23.
1962 *Diorthelasma* cf. *parvum* Cooper; Williams, p. 149, pl. XIII, figs 20–22.

Material. – 49 ventral and 22 dorsal valves. All from blocks 4, 5 and 6, locality A-1230.

Description. – Shell medium-sized, ventri-biconvex in profile and transversely semi-elliptical in outline. Cardinal angles obtuse; maximum width at midvalve. Anterior commissure rectimarginate to weakly sulcate. Ventral valve evenly convex. Interarea apsacline, moderately low, short and curved; delthyrium high and very wide. Dorsal valve weakly convex, with sulcus deeper at the anterior margin. Dorsal interarea low, wide and weakly anacline; notothyrium open, large and very wide. Ornamentation parvicostellate, on some specimens the ribs are distorted. Growth lines variably developed near the anterior margin.

Ventral interior with large deltidiodont teeth supported by large dental plates that converge onto valve floor to form a weak callus on which bilobed diductor muscle scars are impressed. Dorsal interior with a low, thick cardinal process that becomes thicker and more elevated anteriorly. Brachiophores long and narrow. Notothyrial platform on valve floor with crural plates replaced by sockets, or depressions, on valve floor.

Remarks. – The present material is virtually identical to Cooper's type species and thus is assigned to it. In addition, Williams's (1962) species from the Girvan area is here synonmized with the type species because this new material from the White Mountain area, demonstrates that Williams's description is within the range of the type species. Besides, Williams (1962, p. 149) could not find any particular differences between his species and the type species.

Genus *Oanduporella* Hints, 1975

Type species. – *Oanduporella reticulata* Hints, 1975; from the Hirmuse Formation of the Oandu Regional Stage (lower Katian) of Estonia.

Biogeographic distribution. – Lower Sandbian of Bolivia (Havlíček & Branisa 1980; Suárez-Soruco 1992); Sandbian of northern Argentina (Benedetto 1995) and Jones Ridge (Potter & Boucot 1992); upper Sandbian of south-eastern Ireland (Parkes 1992); upper Sandbian to lower Katian of Pomeroy, Ireland (Candela 2003) and Kilbucho, Scotland (Candela & Harper 2010) and lower Katian of Estonia (Hints 1975) and Lithuania (Paškevičius 1994).

Oanduporella kuskokwimensis n. sp.

Plate 13, figures 22–24; Plate 14, figures 1–6

Derivation of name. – Named after the Kuskokwim River, the largest river in the McGrath quadrangle.

Holotype. – MGUH 29633 (Pl. 13, fig. 24; Pl. 14, fig. 3, 4).

Paratypes. – MGUH 29632 (Pl. 13, figs 22, 23; Pl. 14, figs 5, 6) and MGUH 29634 (Pl. 14, figs 1, 2).

Material. – Two pair of conjoined valves, 21 ventral valves and 24 dorsal valves. All from blocks 4–6, locality A-1230.

Diagnosis. – Medium-sized, ventri-biconvex profile with semielliptical–subcircular outline. Cardinal angles obtuse. Rectimarginate to weakly sulcate. Ventral interarea apsacline, short. Dorsal interarea anacline, high.

Delthyrium and notothyrium open, wide. Parvicostellate ornament. Pseudopunctae in interspaces. Ventral interior with large teeth supported by large dental plates. Diductor muscle scars bilobed. Dorsal interior with thick, rod-like cardinal process. Brachiophores long, narrow and supported by brachiophore plates.

Description. – Shell medium-sized, ventri-biconvex in profile and semielliptical–subcircular in outline. Cardinal angles obtuse; maximum width at mid-valve, about 90% as long as wide. Anterior commissure rectimarginate to weakly sulcate. Ventral valve evenly convex. Interarea medium-sized, short and strongly apsacline; delthyrium long and relatively narrow. Dorsal valve weakly convex, with sulcus deeper at the anterior margin. Interarea high, wide and weakly anacline; notothyrium open and wide. Ornamentation parvicostellate with 15 primary costae; up to 40–45 costellae developed at the anterior margin; one growth line variably developed close to the anterior margin. Pseudopunctae in interspaces.

Ventral interior with large teeth supported by large dental plates that converge onto valve floor. Diductor muscle scars bilobed and weakly impressed. Dorsal interior with a very thick, rod-like cardinal process. Brachiophores long, narrow and supported by brachiophore plates extending anteriorly to form an elongate notothyrial platform on valve floor; anteriorly the platform continues as a short, thick median elevation.

Remarks. – This is the first published report of this genus in North America, indicating faunal links to Baltica, though it is also found in Gondwana (Bolivia and Argentina). It possibly indicates a migration during early Sandbian from peri-Gondwana towards the Nixon Fork Subterrane before the genus crossed the Iapetus Ocean (via the Midland Valley Terrane) to reach Baltica in the early Katian. Rasmussen (2011)

summarized the known occurrences of the genus and demonstrated phylogenetic adaptations within the genus that could be achieved through a monophyletic origin in high southerly latitudes. Using palaeo-oceanographical models and palaeogeographic reconstructions, these adaptations were shown to coincide with two dispersal routes: one eastbound via the Northern Precordilera and Avalonia and one westbound via the Farewell Terrane, Jones Ridge, the Midland Valley Terrane, before reaching its final destination in the platform settings of Baltica.

Genus *Salopina* Boucot, 1960

Type species. – *Orthis lunata* Sowerby, 1839; from the Ludlow of England.

Biogeographic distribution. – Reported from lower Sandbian to lower Katian of the Midland Valley Terrane (Williams 1962; Mitchell 1977; Clarkson *et al.* 1992; Candela 2003; Candela & Harper 2010); the upper Katian of Bohemia (Havlíček 1982), the Klamath Terrane (Potter 1990b) and South China (Xu 1996).

***Salopina*? sp.**

Plate 14, figures 7–8

Material. – A pair of conjoined valves and a dorsal valve. All from locality 79WG19.

Description. – Shell small, strongly ventri-biconvex in profile and rectangular to ellipsoideal in outline; weakly bilobate. Ventral valve evenly convex; dorsal valve flat to weakly convex with small sulcus, deeper towards the anterior commissure. Ornamentation consists of even sized, closely spaced costae, though costellae may be developed.

Family Saukrodictyidae Wright, 1964

Genus *Callositella* n. gen.

Derivation of name. – Refers to the characteristic dorsal adductor muscle field that is elevated on callosities.

Type species. – *Callositella cheeneetnukensis* n. sp. from the Sandbian of the White Mountain area, McGrath A-4 and A-5 quadrangles, west-central Alaska.

Diagnosis. – Shell ventri-biconvex, transverse to subquadrate–semi-ellipsoidal in outline. Ventral interarea apsacline, curved, high. Dorsal interarea strongly anacline, high and wide. Ornamentation multicostellate.

Delthyrium wide. Diductor scars elongate, widely separated. Callus depressed; ventral median septum long. Notothyrial platform elevated; blade-like cardinal process; crural plates wide. Adductor muscle field developed on transverse, elevated, callosities.

Description. – Shell ventri-biconvex, transverse to subquadrate–semi-ellipsoidal in outline. Anterior commissure rectimarginate. Ventral interarea apsacline, curved, high and wide. Dorsal interarea strongly anacline, high and wide. Ornamentation multicostellate, with low, rounded costellae.

Ventral interior with a triangular to subcircular wide delthyrium. Diductor scars elongate, widely separated. Callus depressed but extending as broad, elevated ventral median septum almost to the anterior commissure. Teeth large, supported by dental plates converging onto thickened shell interior with strongly impressed pinnate mantle canal system, extending to the anterior commissure. Dorsal interior with highly elevated notothyrial platform; blade-like cardinal process; wide crural plates broadening anteriorly. Notothyrial platform supported by dorsal median septum elevated posteriorly, becoming shallower anteriorly as it fuses with valve floor at mid-valve length. Adductor muscle field developed on transverse, elevated, callosities fusing with anterior end of median septum.

Callositella cheeneetnukensis n. sp.

Plate 14, figures 9–13

Derivation of name. – Named after the Cheeneetnuk River, the nearest large river running northeast–southwest approximately 3 km southeast of the locality (see Fig. 1).

Holotype. – MGUH 29637 (Pl. 14, figs 10, 12).

Paratypes. – MGUH 29636 (Pl. 14, figs 9, 11) and MGUH 29638 (Pl. 14, fig. 13).

Material. – Eight ventral and eight dorsal valves. All from blocks 4–6, locality A-1230.

Diagnosis. – As for genus.

Description. – See description of genus above.

Remarks. – This material is placed within the Saukrodictyidae based on the characteristic adductor muscle platform in the dorsal valve and the shape of the ventral delthyrium. The same adductor muscle platform may explain why Potter & Boucot (1992) reported *Eripnifera* from the White Mountain area.

The current study has not found *Eripnifera* within the White Mountain material. The current material is differentiated from *Eripnifera* because the ventral valves of the present material exhibit a strongly impressed mantle canal system, as well as an extended callus. This is not known from any other genera within the Saukrodictyidae. These differences justify the erection of a new genus.

Family Linoporellidae Schuchert & Cooper, 1931

Genus *Laticrura* Cooper, 1956

Type species. – *Laticrura pionodema* Cooper, 1956 from the lower Sandbian of Virginia, USA (Effna–Rich Valley and Edinburg formations).

Biogeographic distribution. – Lower Sandbian of Virginia (Cooper 1956); Sandbian–upper Katian of the Klamath Mountains (Potter 1990b); upper Sandbian – middle Katian of the Midland Valley Terrane (Williams 1962; Wright 1964; Candela 2006a; Harper 2006; Candela & Harper 2010); lower Katian – lower Cautleyan of Estonia (Hints 1975) and finally reported from the Rawtheyan of Belgium (Sheehan 1987) and Taimyr, northern Russia (Cocks & Modzalevskaya 1997),

Laticrura aff. *pionodema* Cooper, 1956

Plate 14, figures 14–17

1956 *Laticrura pionodema* Cooper new species, p. 983, pl. 144A, figs 1–25, pl. 144C, figs 30–33; pl. 145B, figs 14–16; pl. 146A, figs 1–4.
2010 *Laticrura pionodema* Cooper; Candela & Harper, p. 19, fig. 9p, u.

Material. – Four ventral and seven dorsal valves. All from blocks 4–6, locality A-1230.

Description. – Shell medium-sized, biconvex, probably unisulcate. Ornamentation multicostellate, with concentric growth lines developed in the anterior half of valve. Ventral interarea high, wide and strongly apsacline; delthyrium open, wide and high. Dorsal interarea high, wide and strongly anacline; notothyrium open, wide.

Ventral interior with large teeth supported by strong, divergent dental plates that converge onto valve floor to form elevated callus; accentuated, bilobed diductor muscle impressions separated by thick, elevated septum extending in the anterior part of valve; septum higher on callus than anteriorly.

Dorsal interior with a simple, thin, blade-like cardinal process confined by long and very divergent brachiophores; brachiophores rectangular in section, continuing anterior to the notothyrial platform. Cardinal process variably developed. Crural plates fuse forming a strong, high dorsal median septum that extends anteriorly. Pair of weak side septa extending anteriorly up to mid-valve, is also present.

Remarks. – Although the material is poorly preserved, the valve interiors permit generic determination. The assignment to *L. pionodema* is based on the length of the brachiophores as described by Cooper (1956)

Order Pentamerida Schuchert & Cooper, 1931

Suborder Syntrophiidina Ulrich & Cooper, 1936

Superfamily Camerelloidea Hall and Clarke, 1894

Family Camerellidae Hall and Clarke, 1894

Subfamily Camerellinae Hall and Clarke, 1894

Genus *Camerella* Billings, 1859

Type species. – *Camerella volborthi* Billings, 1859 from the upper Darriwilian Rockland Formation, Ontario, eastern Canada.

Biogeographic distribution. – Known from the Sandbian of eastern Canada and USA (Cooper 1956) and northern Precordillera (Benedetto 2002); Sandbian–lower Katian of the Midland Valley Terrane (Williams 1962; Mitchell 1977; Candela 2006a,b; Candela & Harper 2010); the Katian of Lithuania (Paškevičius 1994), Chinghiz (Klenina *et al.* 1984), Oklahoma (Cooper 1956), Sweden (Sheehan 1979) and the Jones Ridge area, east-central Alaska (Ross & Dutro 1966). In addition it is known from the Middle Llandovery of Girvan (Cocks 2008b).

Camerella sp.

Plate 14, figures 18–19

Material. – Two dorsal valves. All from locality A-1230, blocks 4–6.

Description. – Shell medium-sized with convex dorsal profile, and subcircular to pentagonal outline. Maximum width just posterior to mid-valve. Dorsal fold

with four pronounced costae; ornament of 15 broad costae developed along the anterior margin. Posterior third of valve smooth.

Dorsal interior with narrow, very short and sessile cruralium, confined to umbonal region but continuing as pronounced, thin dorsal median septum within posterior half of valve. Two rounded adductor muscle platforms are developed on both sides of the dorsal median septum.

Remarks. – The present specimens resemble, to some extent, *Parastrophina scotica* Williams (1963). However, they are identified as *Camerella* on the basis of the number of costae in the dorsal fold and the position of the cruralium. The callosities resemble those seen in *Perimecocoelia* Cooper, 1956. However, this taxon differs in outline and ornament (see below).

Genus *Brevicamera* Cooper, 1956

Type species. – *Brevicamera camerata* Cooper, 1956; from the upper Darrivilian Pratt Ferry Formation of Alabama, USA.

Biogeographic distribution. – Previously only recorded from upper Darriwilian of Alabama (Cooper 1956).

Brevicamera? sp.

Plate 14, figures 20–23

Material. – One ventral and one dorsal valve. Both from locality 79WG19.

Description. – Shell medium-sized, dorsi-biconvex in outline. Maximum width at or near the posterior margin. Ventral valve unevenly convex with sulcus developed medially. Dorsal exterior with characteristic high 'rectangular' fold with small depression medially; the anterior commissure markedly uniplicate. Ribs very coarse, developed in anterior half of valve, numbering 14 to 15 on both valves. Two broad costae developed in the dorsal fold. Ventral interior with small, recessive dental plates.

Remarks. – The current occurrence extends the range of the genus from the upper Darriwilian to the Katian. This occurrence may have been reported by Potter & Boucot (1992) as *Eospirigerina* from the White Mountain area.

Genus *Perimecocoelia* Cooper, 1956

Type species. – *Perimecocoelia semicostata* Cooper, 1956; from the upper Darriwilian Effna Formation of eastern USA.

Biogeographic distribution. – This species is reported from the lower Sandbian of Virginia and Tennessee, eastern USA (Cooper 1956) and from the upper Sandbian – lower Katian of Chinghiz, Kazakhstan (Klenina *et al.* 1984).

Perimecocoelia semicostata Cooper, 1956

Plate 14, figure 24; Plate 15, figures 1–3

> 1956 *Perimecocoelia semicostata* Cooper, new species, p. 594, pl. 108E, figs 21–25; pl. 114C, figs 35–38.
> 1984 *Perimecocoelia* aff. *semicostata* Klenina, Nikitin & Popov, p. 142, pl. XIX, figs 3–5.

Material. – Three ventral and four dorsal valves. All from blocks 4–6, locality A-1230.

Description. – Shell medium to large with biconvex outline, subrounded profile and cardinal angles obtuse. Maximum width at mid-valve or within anterior half of valve. Anterior commissure uniplicate. Valves smooth except for a few very broad costae developed within quarter of valve length from the anterior commissure. Ventral palintrope very short, curved; palintrope probably not present in dorsal valve.

Ventral interior with relatively widely divergent teeth supported by a sessile spondylium that narrows considerably towards valve floor, continuing as a short ventral median septum that terminates before mid-valve. Dorsal interior with deep sockets flanking divergent brachiophores, supported by a narrow cruralium that fuses within posterior quarter of valve to form a high, thin dorsal median septum which continues until the anterior margin.

Remarks. – Although these specimens resemble *Parastrophinella*, they are referred to as *Perimecocoelia semicostata* because of the strong similarity in both outline, profile and interior to Cooper's (1956) specimens of this species.

Subfamily Stenocamarinae Cooper, 1956

Genus *Stenocamara* Cooper, 1956

Type species. – *Stenocamara perplexa* Cooper, 1956; from the upper Darriwilian Mosheim Formation, Tennessee, USA.

Biogeographic distribution. – Until the current study only reported from the upper Darriwilian of Tennessee, USA (Cooper 1956) and the Smøla region,

Norway (Harper 1981), as well as from the late Katian of Estonia (Rõõmusoks 1964).

Stenocamara? sp.

Plate 15, figures 4–7

Material. – One ventral valve and one dorsal valve from locality 79WG19.

Description. – Shell small with an equi-biconvex profile and possibly elongate in outline. Maximum width probably close to hinge line. Valves are smooth. Interareas not preserved.

Ventral interior with long, subparallel inner plates that are slightly basomedianly inclined in the most anterior part. Delthyrium obscured by secondary infilling. Dorsal interior with high median septum supporting short, deep and wide septalium.

Remarks. – This material bears a strong resemblance to Cooper's genus *Stenocamara*. However, the material interpreted here as a ventral valve, could in fact also turn out to be a dorsal valve, if what here is described as secondary infill in the delthyrium instead is a rod-like cardinal process. Obviously more material is needed before this can be determined. The posterior part of the dorsal valve, as well as the septalium, does however, indicate that this is referable to *Stenocamara*.

Genus *Eoanastrophia*, Nikiforova & Sapelnikov, 1973

Type species. – *Eoanastrophia antiquata* Nikiforova & Sapelnikov, 1973, from the Katian Chashmankolon and Archalyk beds, Zeravshan Range, Uzbekistan.

Biogeographic distribution. – Reported from the lower Sandbian – upper Katian of Altai (Kulkov & Severgina 1989; Sennikov *et al.* 2008); upper Sandbian – lower Katian of Armorica (Hammann *et al.* 1982; Mélou 1990) and the Chinghiz Terrane (Klenina *et al.* 1984); lower Katian of New South Wales (Percival 2009) and the Katian of the Carnic Alps, Iberia, Perunica and Sardinia (Villas 1985; Leone *et al.* 1991; Boucot *et al.* 2003).

Eoanastrophia? sp.

Plate 15, figures 8–13

Material. – One ventral and dorsal valve from locality 79WG19. One dorsal fragment from locality 79WG126.

Description. – Shell small to medium-sized with a ventri-biconvex profile, and rectangular to transversely ellipsoidal in outline. Maximum width just posterior to mid-valve length. Median depression in both ventral and dorsal valves; ornament of (estimated) 16–18 broad, triangular costae developed along the entire shell. Short ventral palintrope and a short dorsal interarea may be developed.

Ventral interior with small, narrow spondylium raised on very short ventral septum that only supports spondylium most posteriorly. Teeth small. Dorsal interior with inner plates supporting outer plates that are slightly longer; widely divergent. May possess a median ridge that is developed anterior to the dorsal cardinalia.

Remarks. – The spondylium is similar to those taxa referred to the Parallelelasmatidae. However, the current species lack the long, subparallel inner plates in the dorsal valve. It differs from *Camerella* and also *Parastrophina* exteriorly in that the broad costae are not confined to the anterior margin and interiorly in having a very short ventral median septum that does not support the spondylium along its entire length. These internal characters of the ventral valve more resemble those seen in the early virgianids, such as *Brevilamnulella*. Yet, the inner plates in the dorsal valve are longer than what is normally seen in *Brevilamnulella*. Further, a low dorsal median ridge may be developed in the current species. Both are characters that are not virgianid, but instead more resemble that of the Wenlockian pentameroid genus *Vosmiverstum* Breivel & Breivel, 1970, a genus which in fact share several characters with the current species. These include the profile, the costate angular ribbing developed over the entire shell, the short spondylium supported by even shorter median septum and the above mentioned dorsal cardinalia. However, the current species lack the well defined interareas of *Vosmiverstum*. Still, interareas may be weakly developed. Especially in the dorsal valve, it seems like there is a low flat interarea. The ventral valve may also have a very narrow interarea, but this is not possible to determine with certainty in the one ventral valve available for study. If interareas truly are present in both valves in this late Katian species, this would point to a placement within the Virgianidae (Jin *et al.* 2006), which is ancestral to the Pentameroidea. However, more material is required to confirm this taxonomic assignment. Especially as the dorsal valve is fragmented, we can not with certainty erect a new genus that the material clearly warrants. Instead, we have assigned the current species to *Eoanastrophia* based on the profile, outline, number and appearance of ribs, although, internally, as mentioned, the ventral median septum appears very short, and contrary to the type species of *Eoanastrophia*, does not support the spondylium along its entire length and further apparently is lacking the dorsal septalium known from *Eoanastrophia*. Percival (2009) described a ventral valve of an *Eoanastrophia* sp. from the Quondong Limestone, Bowan Park Subgroup of central New South Wales, Australia. Although mistakenly reported as a dorsal valve, it strongly resembles the ventral valve seen in the current material. However, the Australian species also has a longer ventral septum than the Alaskan species.

Family Parallelelasmatidae Cooper, 1956

Genus *Didymelasma* Cooper, 1956

Type species. – *Didymelasma longicrurum* Cooper, 1956; from the lower–middle Sandbian Lebanon Formation, Tennessee, USA.

Biogeographic distribution. – Reported from the Sandbian of Tennessee, USA (Cooper 1956) and the Chu-Ili Range, Kazakhstan (Popov *et al.* 2002). In addition known from the lower Katian of New South Wales, northeastern Australia (Percival 1991).

Didymelasma? sp.

Plate 15, figures 14–15.

Material. – One ventral valve from blocks 4–6, locality A-1230.

Description. – Ventral valve small and convex, about 95% as wide as long. Outline triangular with maximum width at fifth of valve length from anterior margin. Anterior commissure uniplicate, with broad low sulcus and small fold medially. Exterior smooth.
    Interiors not preserved.

Genus *Schizostrophina* Fu, 1982

Type species. – *Schizostrophina margarita* Fu, 1982, from the lower Sandbian Jinhe Formation of Northwest China.

Biogeographic distribution. – Hitherto only reported from the Sandbian of the Chu-Ili Mountains, Kazakhstan (Popov *et al.* 2002) and the Shaanxi Province, northwestern China (Fu 1982).

Schizostrophina? sp.

Plate 15, figures 16–17

Material. – One ventral valve from locality 79WG19.

Description. – Small, strongly convex profile with an elongately triangular outline. Widest point at anterior commissure. Indentation medianly gives the valve an almost bilobed appearance. Smooth, with weak growth lines.

Interior of valve not preserved.

Remarks. – The material is questionably assigned to this exotic genus, based on outline, profile and the characteristic smooth exterior surface only.

Suborder Pentameridina Schuchert & Cooper, 1931

Superfamily Pentameroidea M'Coy, 1844

Family Virgianidae Boucot & Amsden, 1963

Subfamily Mariannaellinae Sapelnikov & Rukavishnikova, 1975

Genus *Brevilamnulella* Amsden, 1974

Type species. – *Clorinda*? *thebesensis* Savage 1913; from the Hirnantian Leemon Formation of Illinois, USA.

Biogeographic distribution. – Reported from the latest Katian (Rawtheyan) of Altai (Severgina 1978), Baltica (Cocks 1982; Rasmussen *et al.* 2010) and the Chinghiz and Chu-Ili terranes (Sapelnikov & Rukavishnikova 1975). From the Hirnantian the genus is known from the District of Mackenzie, northern Canada (Jin & Chatterton 1997), Kolyma (Oradovskaya 1983; Sapelnikov 1985), Illinois, Missouri and Oklahoma (Amsden 1974), South China (Chen *et al.* 2000), Tien-Shan (Menakova 1984). In addition the genus is known from possible Upper Ordovician rocks of North Greenland (Rasmussen 2009).

Brevilamnulella sp.

Plate 15, figures 18–20

Material. – One dorsal valve from locality 79WG19 (marked upper horizon).

Remarks. – Dorsal valve large and strongly convex with smooth exterior. Interarea not preserved. Inner plates extremely abbreviated, outer plates longer and slightly divergent. A median fold may be developed, thus developing a unisulcate anterior commissure.

Description. – Despite only one dorsal valve present in the material, the combination of a smooth exterior

with a dorsal fold and extremely abbreviated inner plates is only seen in *Brevilamnulella*, within the Virgianidae. The current specimen, however, also strongly resembles the recently described genus *Clorilamnulella*, the earliest known clorindoid brachiopod (Rasmussen *et al.* 2010), particularly as it is difficult to assess whether the short inner plates are basomedially divergent, as in the virgianidae, or inclined, as in the clorindidae. Assignment of the current material to *Brevilamnulella* is therefore based on the large size of the shell, compared to the only known species of *Clorilamnulella* from contemporaneous beds, *C. osmundsbergensis* from the Boda Limestone in Sweden (Rasmussen *et al.* 2010) and furthermore because the outer plates appear similar to those in the type species of *Brevilamnulella*, *B. thebesensis* from the Leemon Formation of Illinois, USA (Amsden 1974), and further longer and less thick than those in species of *Clorilamnulella*.

As with the genus *Galeatellina*, described below, the occurrence of *Brevilamnulella* from locality 79WG19, could indicate that deep-water mounds were positioned somewhere on the slope in this part of the Nixon Fork Subterrane in the latest Katian.

Genus *Galeatellina* Sapelnikov & Rukavishnikova, 1976

Type species. – *Galeatellina kajnarensis* Sapelnikov & Rukavishnikova, 1975; from the latest Katian *Holorhynchus giganteus* Zone of the Chinghiz Mountains, eastern Kazakhstan.

Biogeographic distribution. – Hitherto only recorded from the upper Katian (Rawtheyan) of the Chinghiz Terrane (Sapelnikov & Rukavishnikova, 1976).

Galeatellina n. sp.

Plate 15, figures 21–24; Plate 16, figures 1–3

Material. – One ventral and one dorsal valve from locality 79WG126 (marked upper horizon).

Description. – Shell very small, weakly biconvex with outline triangular to rhomboidal. Ventral valve with median fold and high palintrope, dorsal valve sulcate, also with high palintrope. Anterior commissure unisulcate. Maximum width at mid valve or close to anterior margin. Exterior smooth.

Ventral interior probably with a raised, shallow spondylium (fragmented), on an extremely short but thick ventral median septum that continues as a low ridge on valve floor to about mid valve. Dorsal interior with deep triangular depression instead showing the remnants of some very short inner plates. Outer plates may be divergent (fragmented).

Remarks. – The present specimens look interiorly like the unicostae *Drepanorhyncha* described by Cooper (1956). However, as they are unisulcate and smooth as opposed to parasulcate and costate, they are not congeneric. Instead, the interiors are here regarded as typical of the virgianids. The combination of a ventral fold with a dorsal sulcus, combined with the characteristic virgianid cardinalia, is only seen in the extremely rare *Galeatellina*, a very small virgianid which is only known from the upper Katian (Rawtheyan) of the Chingiz Range in Kazakhstan (Sapelnikov & Rukavishnikova 1976). Compared to the type species, *G. kajnarensisz*, the new species is less globose although it does have a remarkably high ventral palintrope compared to most other pre-Silurian virgianids.

One could argue that the material described here as a ventral valve, could be a dorsal valve. However, we believe that in the present specimen, the spondylium is partly fragmented. If instead the structure represented the outer plates, the left one (seen from above) would not bend towards the middle of the valve. In addition, in the most posterior part of the valve, the 'plates' are fused and further supported by a very short, but thick septum.

Galeatellina has previously only been found associated with deep-water carbonate mounds (Sapelnikov & Rukavishnikova 1975). This suggests that it is closely affiliated with *Brevilamnulella*, described above, a virgianid genus that preferred the same settings, but often was positioned on the seaward side of the mounds in more turbulent conditions (Rasmussen *et al.* 2010).

Thus, the occurrence of these two virgianid genera, could signal that mounds were present somewhere on the slope.

Order Rhynchonellida Kuhn, 1949

Superfamily Rhynchotrematoidea Schuchert, 1913

Family Orthorhynchulidae Cooper, 1956

Genus *Orthorhynchuloides* Williams, 1962

Type species. – *Hemithyris nasuta* McCoy, 1852 from the lower Sandbian Craighead Limestone, Girvan, Scotland.

Biogeographic distribution. – Recorded from the lower Sandbian to Hirnantian of Girvan, Scotland (Williams 1962; Harper 2006).

Orthorhynchuloides? sp.

Plate 16, figures 4–7

Material. – One ventral valve, three dorsal valve and 12 undetermined fragments from locality 79WG19.

Description. – Shell large, elongately oval in outline and strongly ventri-biconvex in profile. Probably with deep ventral sulcus, though material is highly deformed. Dorsal fold high, truncated towards the anterior and thus appears tongue-like. Ornamentation consists of large, rounded simple costae widening anteriorly. Interiors unknown.

Remarks. – The specimens collected are identified as *Orthorhynchuloides* on the basis of the ornament, outline, and the deep ventral sulcus characteristic of this genus.

Order Atrypida Rzhonsnitskaia, 1960

Suborder Anazygidina Copper & Gourvennec, 1996

Superfamily Anazygoidea Davidson, 1883

Family Anazygidae Davidson, 1883

Subfamily Anazyginae Davidson, 1883

Genus *Anazyga* Davidson, 1882

Type species. – *Atrypa recurviristra* Hall, 1847; from the early Katian of New York, eastern USA.

Biogeographic distribution. – Reported from the lower Sandbian of Missouri and Pennsylvania (Cooper 1956); upper Sandbian of Tasmania (Laurie 1991); lower Katian of British Columbia, Canada (Jin & Norford 1996) and New York (Patzkowsky & Holland 1997) and also the middle Katian of Hubei, South China (Zhan & Jin 2005). Also known from upper Katian beds of eastern North Greenland (field observations by the first author).

Anazyga sp.

Plate 16, figures 8–10

Material. – One pair of conjoined vales from locality 79WG126.

Description. – Shell small to medium-sized, ventri-biconvex outline and subcircular to elongate profile; maximum width at mid-valve. Ventral umbo sharper, with umbonal zone extending posterior to hinge line. Dorsal valve weakly sulcate. Ornamentation of fine costae, originating in the umbonal zone. Growth lines variably developed more continuously towards the anterior margin. Interiors not preserved.

Genus *Zygospira* Hall, 1862

Type species. – *Producta modesta* Say *in* Hall, 1847; from the upper Katian of Ohio, eastern USA.

Biogeographic distribution. – Sandbian of New South Wales, Australia (Percival 1991); lower Sandbian – lower Katian of Girvan (Williams 1962); lower Katian of Lithuania (Candela 2006a,b); Katian of Indiana, Iowa, Tennessee and Texas, USA (Howe 1988); Rawtheyan of Chinghiz, Kazakhstan (Rong & Boucot 1998), South China and Wales (Zhan & Jin 2005); Hirnantian of the Czech Republic and Norway (Cocks 1988).

Zygospira? sp.

Plate 16, figure 11

Material. – Five fragments, of which three are ventral valves and two are from a dorsal valve. All from locality 79WG19.

Description. – Shell medium-sized, with unevenly convex ventral valve possibly with fold medially. Ornamentation of about 20 simple costae better developed towards the anterior margin.
No interiors preserved.

Subfamily Catazyginae Copper, 1977

Genus *Catazyga* Hall & Clarke, 1893

Type species. – *Athyris headi* Billings, 1862; from the upper Katian of Québec, Canada.

Biogeographic distribution. – Lower Katian of Norway (Neuman *et al.* 1997); lower–middle Katian of Altai-Sayan (Severgina 1978); Katian of Girvan (Williams 1962; Copper 1977; Harper 2006); middle–upper Katian of eastern USA (Patzkowsky & Holland 1997); Cautleyan of Pomeroy (Mitchell 1977); Rawtheyan of Belgium (Sheehan 1987) and Taimyr, northern Russia (Cocks & Modzalevskaya 1997).

Catazyga? sp.

Plate 16, figure 12

Material. – Thirteen fragments, of which four can be identified as ventral and dorsal from locality 79WG19. Also two ventral and three dorsals from locality 79WG126.

Description. – Shell medium-sized, elongately triangular to subrounded. Ventral valve strongly and evenly convex. Ornament of simple rounded, closely spaced costae. Ventral interarea very narrow.

Suborder Lissatrypidina Copper & Gourvennec, 1996

Superfamily Protozygoidea Copper, 1986

Family Cyclospiridae Schuchert, 1913

Genus *Cyclospira* Hall & Clarke, 1893

Type species. – *Cyclospira bisulcata* Emmons, 1842; from the lower Katian Trenton Limestone of New York, eastern USA.

Biogeographic distribution. – Lower Sandbian – upper Katian of the Midland Valley Terrane (Williams 1962; Mitchell 1977; Copper 1986; Harper 2000, 2006; Candela 2003); Laurentia (Emmons 1842; Copper 1986; Neuman 1994; Rong & Zhan 1996); Sandbian–lower Katian of the Chinghiz Range, Kazakhstan (Klenina *et al.* 1984), Kolyma (Rozman 1964; Copper 1986), the Southern Shan States, Myanmar (Cocks & Zhan 1998) and Tien-Shan (Copper 1986); upper Sandbian – lower Katian of northwestern China (Rong *et al.* 1999); lower–middle Katian of Baltica (Copper 1986; Cocks 2005); middle–upper Katian of South China (Zhan & Jin 2005); Rawtheyan of Taimyr, North Russia (Cocks & Modzalevskaya 1997). In addition to the Laurentian occurrences mentioned above, the genus has been reported from Sandbian or Katian rocks of the Jones Ridge, east-central Alaska (Ross & Dutro 1966).

Cyclospira orbus Cocks & Modzalevskaya, 1997

Plate 16, figures 13–19

> 1997 *Cyclospira orbus* sp. nov. Cocks & Modzalevskaya, p. 1082, pl. 5, figs 1, 3–6; Text-figs 5–6.

Material. – 14 pairs of conjoined specimens, 13 ventral valves and 11 dorsal valves. All from locality 79WG19.

Description. – Shell small, ventri-biconvex in outline and oval to rhomboidal in profile. Maximum width at mid-valve. Ventral palintrope orthocline to weakly apsacline. Apical foramen extending beyond posterior margin. Dorsal valve evenly convex and slightly sulcate. Valves smooth, though perforated by minute holes (see SEM image on pl. 16, fig. 19).

Ventral interior with small teeth and small, widely divergent recessive dental plates. Dorsal interior with medianly directed spiralia preserved; low, thick dorsal median septum extending for about 75% of valve length. Relatively deep dental sockets seen posterior to spiralia.

Remarks. – Though smaller than the specimens figured by Cocks & Modzalevskaya (1997), the present specimens are assigned to *C. orbus* on the basis of the oval to rhomboidal outline, the absence of a dorsal fold in the shallow dorsal sulcus and a rather thick and long dorsal median septum. In addition, the dorsal valve is less convex in *C. orbus* than seen in other species of *Cyclospira*. This is also the case with the present specimens. Ross & Dutro (1966) figured *C.* cf. *glansfagea* and *Cyclospira*? sp. from the Jones Ridge area. The latter does resemble the present specimens, but differs in having a wider outline and a much more convex dorsal valve. *C. orbus* indicates a Rawtheyan age for locality 79WG19.

Cyclospira elegantula Rozman, 1964

Plate 16, figures 20–23

> 1964 *Cyclospira*? *elegantula* sp. nov. Rozman, p. 188, pl. 23, figs 4–5.
> 1984 *Cyclospira*(?) *elegantula* Rozman; Klenina, Nikitin & Popov, p. 118, pl. XII, figs 12–17.

Material. – A pair of conjoined valves and a dorsal valve from blocks 4–6, locality A-1230.

Description. – Shell small with strongly ventri-biconvex outline and elongate profile; maximum width at acute cardinal angles. Ventral valve unevenly convex with relatively flat cardinal extremities; median part of valve strongly convex with small sulcus. Dorsal valve weakly convex, with moderate sulcus possessing a fold medianly; anterior commissure uniplicate. Shell smooth, except for minute growth lines seen all over valve surface.

Remarks. – The present specimens resemble *C. glansfagea* Cooper & Kindle, 1936, which also occurs at Jones Ridge. However, they differ in having a deeper ventral valve. This character differentiates the present specimens from most of the other known species of *Cyclospira*, except for the type species *C. bisulcata*

Emmons, from the lower Katian Trenton Limestone of New York. The outline and, notably, the very flat ventral cardinal extremities followed by the sharply defined folds and median sulcus, are identical to those observed in *C. elegantula* Rozman, 1964 from the Sandbian–lower Katian interval of the Selennayakh Range, north-east Russia (Kolyma) and Chinghiz Range of Kazakhstan (Klenina *et al.* 1984). Consequently, the occurrence of this species links the fauna to Kolyma and the Kazakh terranes.

The present specimens further resembles *Jolkinia* Breivel & Breivel, 1988 from the Ural Mountains, Russia. However, this genus is so far only recorded from the Wenlock, and thus, is less likely to be found in Sandbian strata.

Order Unknown
Family Unknown

Fam., gen. *et* sp. indet. 1

Plate 16, figure 24; Plate 17, figures 1–3

Material. – One ventral valve from blocks 4-6, locality A-1230, and one ventral valve from locality 79WG126.

Description. – Ventral valve small, circular and strongly convex. Maximum convexity in anterior half of valve. Semi-oval to rectangular in outline. Interarea not preserved. Ornamentation consists of widely spaced, simple, blade-like costae. Valve strongly lamellose in its anterior one-fourth. May have a weak depression medianly.

Ventral interior with elevated callus, and deeply impressed V-shaped diductor muscle scars.

Remarks. – The deep diductor muscle grooves resemble what is seen in *Glyptorthis*. However, shell profile and outline is too different from that genus to be referred to it.

Fam., gen. *et* sp. indet. 2

Plate 17, figures 4–6

Material. – One ventral valve from locality 79WG126.

Description. – Shell small, strongly pyramidal profile. Outline not possible to deduce, although appear widest towards the anterior half. Ornament of closely spaced rounded costae. Interarea short, high and curved. Teeth large. Delthyrium open.

Ventral interior with strong parallel dental plates.

Fam., gen. *et* sp. indet. 3

Plate 17, figures 7–8

Material. – One ventral valve from locality 79WG126.

Description. – Small, slightly convex profile with a strongly alate and bilobed outline. Interarea small, curved. Delthyrium open, wide. Teeth low, broad. Smooth exterior. Interior with a sessile spondylium.

Remarks. – This material could be an extremely deteriorated shell of *Duolobella*, where the costate ornament is not preserved. The bilobed outline, the configuration of the characteristic wide, low teeth and the elevated callus are all typical of *Duolobella*. However, although closely affiliated with this genus, the outline is very different in this species being much narrower, much wider than long. In addition, the ventral interarea is much smaller and with a relatively wider delthyrium. Further, in contrast to the lower Sandbian *Duolobella*, the current species is only found at the Katian locality 79WG126, also suggesting that this is indeed a separate genus.

Fam., gen. *et* sp. indet. 4

Plate 17, figures 9–11

Material. – One ventral valve from locality 79WG19.

Description. – Small, strongly convex in ventral profile and transverse to subcircular or subquadrate in outline. Maximum width at mid-valve; anterior commissure rectimarginate. Ventral valve unevenly convex, almost triangular to subpyramidal with marked umbo posterior to hinge line. Interarea probably not present. Delthyrium high and wide. Shell is smooth.

No clear structures are visible in the interior of the valve.

Fam., gen. *et* sp. indet. 5

Plate 17, figures 12–13

Material. – One ventral? valve from locality 79WG19.

Description. – Small to medium-sized, strongly convex profile and with a transversely triangular outline of the posterior part of the shell. Anterior region not preserved. Maximum width probably at the hingeline marked by cardinal extremities. Ventral valve deep with a rounded to pyramidal apex. Umbo positioned approximately at the strophic hinge line. Interarea may be high and narrow, but could also be interior

structures. Valve is smooth, but with one or two concentric growth lines.

Interior with large shallow, bilobed muscle scar depression occupying most of the space in the fragmented specimen.

Fam., gen. *et* sp. indet. 6

Plate 17, figures 14–16

Material. – One conjoined specimen from locality 79WG126.

Description. – Small, planoconvex profile with highest convexity medially where a fold is developed in the ventral valve. Outline transversely sub-rounded, possibly alate. Widest point at hinge line. Cardinal extremities acute. The ventral fold consists of two or possibly three smaller folds with one or two indentations in between. Ventral interarea wide, apsacline, slightly curved. Dorsal valve flat, although may poses a shallow, wide sulcus medially with two indentations corresponding to those in the ventral valve. Ornament of closely spaced costae that all originate in the umbonal region. Dorsal interarea anacline to orthocline, equalsized with its ventral counterpart.

Remarks. – The outline of this species initially suggests assignment to the plectambonitoids, perhaps even a very small specimen of *Xenambonites* based on the ornament. However, the strong ventral fold with its indentations, as well as the lack of a ventrally directed geniculation suggest that this specimen belongs to a separate genus. Further, this shell is from the much younger upper Katian locality 79WG126, whereas *Xenambonites* is confined to the lower Sandbian. Could also in outline and ornament resemble an early spiriferoid. However, as the interiors are not really observable, this is impossible to determine based on the present material.

Fam., gen. *et* sp. indet. 7

Plate 17, figures 17–20

Material. – One ventral and one dorsal valve. Both from locality 79WG19.

Description. – Small-sized, planoconvex profile with a suboval to pentagonal outline. Maximum width at the acute cardinal extremities. Ventral valve unevenly convex with a rounded, convex profile. Fold developed medially as is two coarse costae on the otherwise smooth shell. Dorsal exterior more pentagonal in outline with a clear strophic appearance. Characterized by what is probably a deep sulcus with two indentations corresponding to the ventral costae, so that the sulcus consist of one large costa. Anterior commissure markedly uniplicate. No interareas or palintropes observed.

Ventral interior not preserved. Dorsal interior with a pair of small, divergent plates.

Acknowledgements. – Sten Lennart Jakobsen is thanked for helping with the conservation of the specimens. Jisuo Jin, London, Ontario, is thanked for many constructive comments and input that helped refine the taxonomy. Furthermore, we thank Yves Candela, Edinburgh, and Ren-bin Zhan, Nanjing, for their constructive reviews. CMØR acknowledges the Danish National Research Foundation for support to the Center for Macroecology, Evolution and Climate, as well as support to the Nordic Center for Earth Evolution (NordCEE). DATH acknowledges support from the Danish Council for Independent Research (FNU), who also funded the scanning electron microscope. Finally, we would like to thank Wyatt Gibert and Alfred Potter, who together with RBB, did the fieldwork in the late seventies. Al Potter is further acknowledged for the tremendous work he carried out in preparing this material for study. This work would not have been initiated or completed without the vision and support of Arthur (Art) J. Boucot. This paper is a contribution to IGCP project 591 'The Early to Middle Paleozoic revolution'.

References

Adrain, J.M., Chatterton, B.D.E. & Blodgett, R.B. 1995: Silurian trilobites from southwestern Alaska. *Journal of Paleontology 69*, 723–736.

Alberstadt, L.P. 1973: Articulate brachiopods of the Viola Formation (Ordovician) in the Arbuckle Mountains, Oklahoma. *Oklahoma Geological Survey Bulletin 117*, 1–89.

Amsden, T.W. 1974: Late Ordovician and Early Silurian articulate brachiopods from Oklahoma, southwestern Illinois, and eastern Missouri. *Oklahoma Geological Survey Bulletin 119*, 1–154.

Andreeva, O.N. & Nikiforova, O.I. 1955: Brakiopody [Klass Brachiopoda], 61–81. *In* Nikiforova, O.I. (ed.): Poevoi atlas ordovikskoi i siluriiskoi fauny sibirskoi platformy [Field atlas of Ordovician and Silurian faunas of the Siberian Platform], 268. VESGEI, Gosgeoltekhizdat, Moskva, [in Russian].

Angelin, N.P. & Lindström, G. 1880: Fragmenta Silurica e dono Caroli Henrici Wegelin. *Stockholm*, 60 pp.

Baarli, B.G. 1988: The Llandovery enteletacean brachiopods of the central Oslo region, Norway. *Palaeontology 31*, 1101–1129.

Baarli, B.G. 1995: Orthacean and strophomenid brachiopods from the Lower Silurian of the central Oslo Region. *Fossils and Strata 39*, 93 pp.

Baarli, B.G. & Harper, D.A.T. 1986: Relict Ordovician brachiopod faunas in the Lower Silurian of Asker, Oslo Region, Norway. *Norsk Geologisk Tidsskrift 66*, 87–98.

Barrande, J. 1879: Iére partie. Recherches paléontologiques, vol. 5. Classe de Mollusques. *In: Système Silurien du centre de la Bohême*, 226. Published by the author, Prague & Paris.

Bekker, H. 1921: The Kuckers Stage of the Ordovician rocks of north-east Estonia. *Acta et commentationes.Universitatis Tartuensis (Dorpatensis) A2*, 1–92.

Benedetto, J.L. 1995: La fauna de braquiópodos de la formación Las Plantas (Ordovicico Tardío, Caradoc), Precordillera Argentina. *Revista Española de Paleontología 10*, 239–258.

Benedetto, J.L. 2002: The Ordovician brachiopod faunas of Argentina: chronology and biostratigraphic succession. *Correlación Geológica 16*, 87–106.

Bergström, J. 1968: Upper Ordovician brachiopods from Västergötland, Sweden. *Geologica et palaeontologica 2*, 1–35.

Billings, E. 1857: Report of progress years 1856. *Geological Survey of Canada, Report of progress* for the years 1853–54–55–56. 247–345.

Billings, E. 1859: Description of a new genus of Brachiopoda, and on the genus Cyrtodonta. *Canadian Naturalist and Geologist 4*, 301–303.

Billings, E. 1862: New species of fossils from different parts of the lower, middle, and upper Silurian rocks of Canada. 96–185. *Palaeozoic Fossils, volume 1*, no. 4. Geological Survey of Canada, Montreal, 426 pp.

Blodgett, R.B. & Boucot, A.J. 1999: Late Early Devonian (late Emsian) eospiriferinid brachiopods from Shellabarger Pass, Talkeetna C-6 quadrangle, south-central Alaska and their biogeographic importance; further evidence for a Siberian origin of the Farewell and allied Alaskan terranes. *Senckenbergiana lethaea 72*, 209–221.

Blodgett, R.B. & Clough, J.G. 1985: The Nixon Fork terrane – part of an *in-situ* peninsula extension of the Paleozoic North American continent. *Geological Society of America, Abstracts with Programs 17*, 342.

Blodgett, R.B., Rohr, D.M. & Boucot, A.J. 2002: Paleozoic linkages between Alaskan accreted terranes and Siberia, 273–291. *In* Miller, E.L., Grantz, A. & Klemperer, S.L. (eds): *Tectonic Evolution of the Bering Shelf – Chukchi Sea-Arctic Margin and Adjacent Landmasses*. Geological Society of America Special Paper, *360*, Boulder, Colorado, 387 pp.

Blodgett, R.B., Boucot, A.J., Rohr, D.M. & Pedder, A.E.H. 2010: The Alexander Terrane – a displaced fragment of Northeast Russia? – Evidence from Silurian–Middle Devonian megafossils and stratigraphy. *Memoirs of the Association of Australasian Palaeontologists 39*, 325–341.

Borissiak, M.A. 1956: Rod *Kassinella* [genus *Kassinella*]. Materiali vesesoyuznogo nauchnoissledovatelskii Geologisckeskova Instituta, Moscow (new series) *12*, 50–52. [in Russian]

Botquelen, A. & Mélou, M. 2007: Caradoc brachiopods from the Armorican Massif (Northwestern France). *Journal of Paleontology 81*, 1080–1090.

Boucot, A.J. 1960: Brachiopoda. 2–14, *In* Boucot, A.J., Martinsson, R., Thorsteinsson, R., Walliser, O.H., Whittington, H.B. & Yochelson, E.L. (eds): *A Late Silurian fauna from the Sutherland River Formation, Devon Island, Canadian Archipelago*. Bulletin of the Geological Survey of Canada 65, 51 pp.

Boucot, A.J. 1975: *Evolution and extinction rate controls: Developments in Palaeontology and Stratigraphy, 1*. Elsevier Scientific Publishing Company, Amsterdam, 427.

Boucot, A.J. & Amsden, T.W. 1963: Virgianidae, a new family of pentameracean brachiopods. *Journal of Paleontology 37*, 296.

Boucot, A.J. & Johnson, J.G. 1967: Silurian and Upper Ordovician atrypids of the genera *Plectatrypa* and *Spirigerina*. *Norsk geologisk Tidsskrift 47*, 79–101.

Boucot, A.J., Rong, J.-y., Chen, X. & Scotese, C.R. 2003: Pre-Hirnantian Ashgill climatically warm event in the Mediterranean region. *Lethaia 36*, 119–132.

Brabb, E.E. 1967: Stratigraphy of the Cambrian and Ordovician rocks in east-central Alaska: lower Paleozoic paleontology and stratigraphy of east-central Alaska. *Geological Survey Professional Paper 559-A*, 30.

Breivel, I.A. & Breivel, M.G. 1970: Dva siluriyskikh roda iz nadsemeystva Pentameracea [Two Silurian genera of the superfamily Pentameracea], 52–56. *In* Breivel, M.G. & Papulov, G.N. (eds): *Materialy po Geologii Urala [Materials on the geology of the Urals]*. UFIGiG, AN SSSR, Sverdlovsk. [in Russian].

Breivel, I.A. & Breivel, M.G. 1988: Biostratigrafiia i Brahhiopody Silura Vostochnogo Sklona Urala [Biostratigraphy and brachiopods of the Silurian of the eastern slopes of the Urals]. *Ministervo Geologii SSSR, Uralskoe proizvodstvennoe geologicheskoe otdelenie*. Nedra, Moskva, 204 pp, 57 pl. [in Russian].

Candela, Y. 2003: *Late Ordovician brachiopods from the Bardahessiagh Formation of Pomeroy, Ireland*. Monograph of the Palaeontological Society, London, (Publ. No. 618, part of Vol. 156 for 2002), 95, 12 pls.

Candela, Y. 2006a: Statistical comparisons of late Caradoc (Ordovician) brachiopod faunas around the Iapetus Ocean, and terranes located around Australia, Kazakhstan and China. *Geodiversitas 28*, 433–446.

Candela, Y. 2006b: Late Ordovician brachiopod faunas from Pomeroy, Northern Ireland: a palaeoenvironmental synthesis. *Transactions of the Royal Society of Edinburgh: Earth Sciences 96*, 317–325.

Candela, Y. & Harper, D.A.T. 2010: Late Ordovician (Katian) brachiopods from the Southern Uplands of Scotland: biogeographic patterns on the edge of Laurentia. *Earth and Environmental Science Transactions of the Royal Society of Edinburgh 100*, 253–274.

Carlisle, H. 1979: Ordovician stratigraphy of the Tramore area, County Waterford, with a revised Ordovician correlation for southeast Ireland, 545–554. *In* Harris, A.L., Holland, C.H. & Leake, B.E. (eds.): *The Caledonides of the British Isles – reviewed.* Geological Society of London, Special Publications, 8, Scottish Academic Press, Edinburgh, 768 pp.

Chang, M.-l. 1983: Brachiopods from the Miaopo Formation in Huanghuachang of Yichang, Hubei Province. *Acta palaeontologica Sinica 22*, 474–481.

Chen, X., Rong, J.-y., Mitchell, C.E., Harper, D.A.T., Fan, J., Zhan, R.-b., Zhang, Y., Li, R.-y. & Wang, Y. 2000: Late Ordovician to earliest Silurian graptolite and brachiopod biozonation from the Yangtze region, South China, with a global correlation. *Geological Magazine 137*, 623–650.

Chen, P., Jin, J. & Lenz, A.C. 2008: Evolution, palaeoecology, and palaeobiogeography of the Late Ordovician – Early Silurian brachiopod *Epitomyonia*. *Palaeoworld 17*, 85–101.

Chiang, K.K. 1972: *Hesperorthis* and two new Silurian species. *Journal of Paleontology 46*, 353–359.

Churkin, Jr. M. & Carter, C. 1970: Early Silurian Graptolites from southeastern Alaska and their correlation with graptolitic sequences in North America and the Arctic. *Geological Survey Professional Paper 653*, 51 pp. 4 pls.

Clarkson, E.N.K., Harper, D.A.T., Owen, A.W. & Taylor, C.M. 1992: Ordovician faunas in mass-flow deposits, Southern Scotland. *Terra Nova 4*, 245–253.

Cloud, P.E.J. 1948: *Dicaelosia* versus *Bilobites*. *Journal of Paleontology 22*, 373–374.

Cocks, L.R.M. 1970: The Silurian brachiopods of the superfamily Plectambonitacea. *Bulletin of The British Museum (Natural History), Geology 19*, 139–203.

Cocks, L.R.M. 1982: The commoner brachiopods of the latest Ordovician of the Oslo–Asker District, Norway. *Palaeontology 25*, 755–781.

Cocks, L.R.M. 1988: Brachiopods across the Ordovician–Silurian boundary. *Bulletin of the British Museum (Natural History), Geology Series 43*, 311–315.

Cocks, L.R.M. 2005: Strophomenate brachiopods from the late Ordovician Boda Limestone of Sweden: their systematics and implications for palaeogeography. *Journal of Systematic Palaeontology 3*, 243–282.

Cocks, L.R.M. 2008a: *A Revised Review of British Palaeozoic Brachiopods.* Monograph of the Palaeontographical Society London, 1–276 pls. 1–10. (Publ. No. 629, part of vol. 161 for 2007).

Cocks, L.R.M. 2008b: The Middle Llandovery brachiopod fauna of the Newlands Formation, Girvan Scotland. *Journal of Systematic Palaeontology 6*, 61–100.

Cocks, L.R.M. 2010: Caradoc strophomenoid and plectambonitoid brachiopods from Wales and the Welsh Borderland. *Palaeontology 53*, 1155–1200.

Cocks, L.R.M. & Cooper, R.A. 2004: Late Ordovician (Hirnantian) shelly fossils from New Zealand and their significance. *New Zealand Journal of Geology & Geophysics 47*, 71–80.

Cocks, L.R.M. & Fortey, R.A. 1990: Biogeography of Ordovician and Silurian faunas. *Geological Society Memoir 12*, 97–104.

Cocks, L.R.M. & Modzalevskaya, T.L. 1997: Late Ordovician brachiopods from Taimyr, Arctic Russia, and their palaeogeographical significance. *Palaeontology 40*, 1061–1093.

Cocks, L.R.M. & Rong, J.-y. 1988: A review of the Late Ordovician *Foliomena* brachiopod fauna with new data from China, Wales and Poland. *Palaeontology 31*, 53–67.

Cocks, L.R.M. & Rong, J.-y. 1989: Classification and review of the brachiopod superfamily Plectambonitacea. *Bulletin of the British Museum (Natural History), Geology Series 45*, 77–163.

Cocks, L.R.M. & Rong, J.-y. 2000: Strophomenida, 216- 349. *In* Kaesler R.L. (ed.): *Treatise on invertebrate palaeontology: Part H, revised.* Brachiopoda, 2, Geological Society of America, and The University of Kansas Press, Boulder, Colorado & Lawrence, Kansas. 919 pp.

Cocks, L.R.M. & Torsvik, T.H. 2002: Earth geography from 500 to 400 million years ago: a faunal and palaeomagnetic review. *Journal of the Geological Society (London) 159*, 631–644.

Cocks, L.R.M. & Torsvik, T.H. 2007: Siberia, the wandering northern terrane, and its changing geography through the Palaeozoic. *Earth-Science Reviews 82*, 29–74.

Cocks, L.R.M. & Torsvik, T.H. 2011: The Palaeozoic geography of Laurentia and western Laurussia: a stable craton with mobile margins. *Earth-Science Reviews 106*, 1–51.

Cocks, L.R.M. & Zhan, R.-b. 1998: Caradoc brachiopods from the Shan States, Burma (Myanmar). *Bulletin of the British Museum (Natural History), Geology Series 54*, 109–130.

Cooper, G.A. 1930: The brachiopod genus *Pionodema* and its homeomorphs. *Journal of Paleontology 4*, 369–382.

Cooper, G.A. 1956: Chazyan and related brachiopods. *Smithsonian Miscellaneous Collections 127*, 1–1245, pls. 1–269.

Cooper, G.A. & Kindle, C.H. 1936: New brachiopods and trilobites from the Upper Ordovician of Percé, Quebec. *Journal of Paleontology 10*, 348–372.

Copper, P. 1977: *Zygospira* and some related Ordovician and Silurian atrypoid brachiopods. *Palaeontology 20*, 295–335.

Copper, P. 1986: Evolution of the earliest smooth spire-bearing atrypoids (Brachiopoda: Lissatrypidae, Ordovician–Silurian). *Palaeontology, 29*, 827–866.

Copper, P. & Gourvennec, R. 1996: Evolution of the spire-bearing brachiopods (Ordovician–Silurian), 81–88. *In* Copper, P. & Jin, J. (eds.) *Proceedings of the Brachiopods: Proceedings of the Third International Brachiopod Congress.* CRC Press, Sudbury, Ontario, Canada, 2–5 September 1995. 356 pp.

Dahlqvist, P., Harper, D.A.T. & Wickström, L.M. 2007: Late Ordovician shelly faunas from Jämtland, central Sweden, p. 82. *In* Ebbestad, J.O.R., Wickström, L.M. & Högström, A.E.S. (eds): *WOGOGOB 2007, Field guide & abstracts, volume.* Geological Survey of Sweden. Rapporter och meddelanden 128. 110 pp.

Dalman, J.W. 1828: Upställning och Beskrifning af de i Sverige funne Terebratuliter. *Kongliga Vetenskaps-akademien Handlingar för år 1827*, 85–155.

Davidson, T. 1847: Observations on some Wenlock-limestone Brachiopoda, with descriptions of several new species. *London Geological Journal 1*, 52–65.

Davidson, T. 1882: *Supplement to the British Devonian Brachiopoda, volume 5.* Palaeontological Society Monograph, London, 1–62.

Davidson, T. 1883: *A monograph of the British Fossil Brachiopoda, vol. V. Silurian Supplement, part 2.* Palaeontographical Society, London, 135–242.

Decker, J., Bergman, S.C., Blodgett, R.B., Box, S.E., Bundtzen, T.K., Clough, J.G., Coonrad, W.L., Gilbert, W., Miller, M.L., Murphy, J.M., Robinson, M.S. & Wallace, W.K. 1994: Geology of southwestern Alaska, 285–310. *In* Plafker, G. & Berg, H.C. (eds): *The Geology of Alaska. Boulder, Colorado, Geological Society of America, The Geology of North America, G-1.* 1055 pp.

Dewing, K. 1999: Late Ordovician and Early Silurian strophomenid brachiopods of Anticosti Island, Quebéc, Canada. *Palaeontographica Canadiana 17*, 1–143.

Dewing, K. 2004: Shell structure and its bearing on the phylogeny of Late Ordovician -Early Silurian strophomenoid brachiopods from Anticosti Island, Quebec. *Journal of Paleontology 78*, 275–286.

Duméril, A.M.C. 1806: *Zoologie Analytique ou Methode Naturella de Classification des Animaux.* Allais, Paris, 147.

Dumoulin, J.A., Bradley, D.C., Harris, A.G. & Repetski, J.E. 1998: Lower Paleozoic deep-water facies of the Medfra area, central Alaska. *U.S. Geological Survey Professional Paper 1614*, 73–103.

Dumoulin, J.A., Harris, A.G, Bradley, D.C. & de Freitas, T.A. 2000: Facies patterns and conodont biogeography in Arctic Alaska and the Canadian Arctic islands: evidence against juxtaposition of these areas during early Paleozoic time. *Polarforschung 68*, 257–266.

Dutro, J.T. & Ross, R.J. 1964: Probable Late Ordovician (Ashgill) brachiopods from east-central Alaska. *Geological Society of America Special Paper 76*, 53.

Emmons, E. 1842: *Survey of the Second Geological District*, x + 437. White and Visscher, Albany.

Ettensohn, F.R. 2008: The Appalachian foreland basin in eastern United States, 105–181. *In* Miall, A.D. (ed.): *Sedimentary Basins of the World, vol. 5: The Sedimentary Basins of the United States and Canada*. Elsevier B.V., Xiii + 593 pp.

Foerste, A.F. 1909: *Preliminary notes on Cincinnatian fossils*. Bulletin of the Scientific Laboratories of Denison University, *14*, 209–232.

Foerste, A.F. 1914: *Notes on the Lorranie faunas of New York and the provincce of Quebec*. Bulletin of the Denison University Science Laboratories, *17*, 247–328.

Fortey, R.A. & Cocks, L.R.M. 1998: Biogeography and palaeogeography of the Sibumasu terrane in the Ordovician: a review. 43–56. *In* Hall, R. & Holloway, J.D. (eds): *Biogeography and Geological Evolution of SE Asia*, 417. Backbuys Publishers, Leiden.

Fortey, R.A. & Cocks, L.R.M. 2003: Palaeontological evidence bearing on global Ordovician– Silurian continental reconstructions. *Earth-Science Reviews 61*, 245–307.

Fu, L.-p. 1982: Brachiopoda. *In* Xian Institute of Geology and Mineral Resources (ed.): *Paleontological Atlas of Northwest China, Shaanxi-Gansu-Ningxia Volume, Part 1, Precambrian and Early Paleozoic*. Geological Publishing House, Beijing, 95–178, pls. 30–45.

Gilbert, W.G. 1981: Preliminary geologic map and geochemical data, Cheeneetnuk River area, Alaska. *Alaska Open-file Report 153*, 1–10, 12 sheets, scale 11:63.360, Department of Natural Resources, Division of Geological and Geophysical Surveys.

Gorjansky, V.J. & Popov, L.E. 1985: Morfologiia, systematicheskoe polozhenie i proiskhozhdenie bezzamkovykh brakhiopods karbonatnoi rakovinoi [Morphology, systematic position and origin of the inarticulate brachiopods with calcareous shells]. *Paleontologicheskii Zhurnal 1985*, 3.

Gray, J.E. 1840: *Synopsis of the contents of the British Museum*, 42nd edition. British Museum, London, 370.

Hall, J. 1847: *Paleontology of New Yok, vol. 1. Containing descriptions of the organic remains of the lower division of the New-York System*. C. van Benthuysen, New York, 338.

Hall, J. 1859: *Observations on genera of Brachiopoda. Contributions to the palaeontology of New-York*. New York State Cabinet of Natural History, 12th Annual Report, Albany, 8–110.

Hall, J. 1862: *Contributions to Paleontology Comprising Descriptions of New Species of Fossils from the Upper Helderberg and Chemung Groups*. New York State Cabinet of Natural History, 12th Annual Report, Albany, 27–197.

Hall, J. 1883: *Brachiopoda, Plates and Explanations*. New York State Geologist, 2nd Annual Report for 1882, Albany, 17 pp.

Hall, J. & Clarke, J.M. 1892: *An introduction to the study of the genera of Paleozoic Brachiopoda New York Geological Survey*. Charles van Benthuysen & Sons, Albany, 367 pp.

Hall, J. & Clarke, J.M. 1893: *An Introduction to the Study of the genera of Palaeozoic Brachiopoda. Palaeontology of New York, vol. 8, Part 2*. Charles van Benthuysen & Sons. Albany. p. 1–317.

Hall, J. & Clarke, J.M. 1894 [1895]: *An Introduction to the Study of the Genera of Paleozoic Brachiopoda, volume 8*, New York Geological Survey, xvi + 394 pls. 21–84. Charles van Benthuysen & Sons, Albany.

Hallam, A. 1984: Prequarternary sea level changes. *Annual Review of Earth and Planetary Sciences 12*, 205–243.

Hallam, A. 1992: *Phanerozoic Sea-Level Changes*. Columbia University Press, New York, 266 pp.

Hammann, W., Robardet, M. & Romano, M. 1982: *The Ordovician System in Southwestern Europe (France, Spain, and Portugal)*, IGCP Publication 11, 47 pp.

Hammer, Ø. & Harper, D.A.T. 2006: *Palaeontological Data Analysis*, 1st edition, Blackwell Publishing, Oxford, 368 pp.

Hammer, Ø., Harper, D.A.T. & Ryan, P.D. 2001: PAST: Paleontological Statistics Software Package for Education and Data Analysis. *Palaeontologia Electronica 4*, 9 pp.

Hansen, J. 2008: Upper Ordovician brachiopods from the Arnestad and Frognerkilen formations in the Oslo–Asker district, Norway. *Palaeontos 13*, 1–99, 12 pls.

Harper, D.A.T. 1981: The brachiopod Stenocamara from the Ordovician of Smøla, Norway. *Norsk Geologisk Tidsskrift 61*, 149–152.

Harper, D.A.T. 1986: The brachiopod *Ptychopleurella lapworthi* (Davidson) from the Ordovician of Girvan, S.W. Scotland. *Journal of Paleontology 60*, 845–850.

Harper, D.A.T. 1989: Brachiopods from the Upper Ardmillan succession (Ordovician) of the Girvan District, Scotland, Part 2. *Monograph of the Palaeontological Society 142*(579), 79–128.

Harper, D.A.T. 2000: Late Ordovician brachiopod biofacies of the Girvan district, SW Scotland. *Transactions of the Royal Society of Edinburgh: Earth Sciences 91*, 471–477.

Harper, D.A.T. 2006: *Brachiopods from the Upper Ardmillan Succession (Ordovician) of the Girvan district, Scotland. Part 3*. Monograph of the Palaeontographical Society, London, Publ. No. 624, part of Vol. 159 for 2005, 129–187, pls 123–133.

Harper, D.A.T. & Brenchley, P.J. 1993: An endemic brachiopod fauna from the Middle Ordovician of North Wales. *Geological Journal 28*, 21–36.

Harper, D.A.T., Owen, A.W. & Williams, H. 1984: The Middle Ordovician of the Oslo Region, Norway, 34. The type Nakkholmen Formation (upper Caradoc), Oslo, and its faunal significance. *Norsk Geologisk Tidsskrift 64*, 293–312.

Harper, D.A.T., Mac Niocaill, C. & Williams, S.H. 1996: The palaeogeography of the early Ordovician Iapetus terranes: an integration of faunal and palaeomagnetic constraints. *Palaeogeography, Palaeoclimatology, palaeoecology 121*, 297–312.

Havlíček, V. 1950: Ramenonozci Ceského Ordoviku [The Ordovician Brachiopoda from Bohemia]. *Rozpravy Ústredního ústavu geologického 13*, 1–72.

Havlíček, V. 1961: Plectambonitacea im böhmischen Palaözoikum (Brachiopoda). *Věstník Ústředního ústavu geologického 36*, 447–451.

Havlíček, V. 1967: Brachiopoda of the Suborder Strophomenidina in Czechoslovakia. *Rozpravy Ústredního ústavu geologického 33*, 1–236, pls. 231–252.

Havlíček, V. 1968: New brachiopods from the lower Caradoc of Bohemia. *Věstník Ústředního ústavu geologického 43*, 123–125.

Havlíček, V. 1977: Brachiopods of the order Orthida in Czechoslovakia. *Rozpravy Ústredního ústavu geologického 44*, 1–327.

Havlíček, V. 1982: Ordovician in Bohemia: development of the Prague Basin and its benthic communities. *Sbornik geologickych ved Geologie 37*, 103–136.

Havlíček, V. & Branisa, L. 1980: Ordovician brachiopods of Bolivia. *Rozpravy Československé Akademie Věd. Řada Matematických a Přírodních Věd 90*, 1–54.

Havlíček, V. & Mergl, M. 1982: Deep water shelly fauna in the latest Kralodvorian (upper Ordovician, Bohemia). *Věstník Ústředního ústavu geologického 57*, 37–46.

Hiller, N. 1980: Ashgill Brachiopoda from the Glyn Ceiriog District, north Wales. *Bulletin of the British Museum (Natural History), Geology 34*, 109–216.

Hints, L. 1975: *Brakhiopody Enteletacea Ordovika Pribaltiki*. [Ordovician Brachiopods Enteletacea of East Baltic]. Eesti NSV Teaduste Akadeemia Gedoogia Institute, Tallinn, 117 pp [in Russian].

Holloway, J.D. 2004: The trilobite subfamily Monorakinae (Pterygometopidae). *Palaeontology 47*, 1015–1036.

Howe, H.J. 1966: Orthacea from the Montoya Group (Ordovician) of Trans-Pecos Texas. *Journal of Paleontology 40*, 241–257.

Howe, H.J. 1988: Articulate brachiopods from the Richmondian of Tennessee. *Journal of Paleontology 62*, 204–218.

Ingham, J.K. 2000: Scotland: the Midland Valley Terrane – Girvan, 43–47. *In* Fortey, R.A., Harper, D.A.T., Ingham, J.K, Owen, A.W, Rushton, A.W.A. & Woodcock, N.P. (eds): *A Revised Correlation the Ordovician Rocks of the British Isles*. Geological Society, Special Report, *24*, 83 pp.

Jaanusson, V. 1962: *Two Plectambonitacean brachiopods from the Dalby Limestone (Ordovician) of Sweden.* Bulletin of the Geological Institution of Uppsala University, *39*, 1–8.

Jin, J. & Chatterton, B.D.E. 1997: Latest Ordovician–Silurian articulate brachiopods and biostratigraphy of the Avalanche Lake area, southwestern district of Mackenzie, Canada. *Palaeontographica Canadiana 13*, 1–167.

Jin, J. & Copper, P. 1999: The deep–water brachiopod Dicoelosia King, 1850, from the Early Silurian tropical carbonate shelf of Anticosti Island, eastern Canada. *Journal of Paleontology 73*, 1042–1055.

Jin, J. & Norford, B.S. 1996: Upper Middle Ordovician (Caradoc) brachiopods from the Advance Formation, northern Rocky Mountains, British Columbia. *Geological Survey of Canada. Bulletin 491*, 20–77.

Jin, J., Caldwell, W.G.E. & Norford, B.S. 1997: Late Ordovician brachiopods and biostratigraphy of the Hudson Bay Lowlands, northern Manitoba and Ontario. Geological Survey of Canada. *Bulletin 513*, 1–115.

Jin, J. & Zhan, R.-b. 2000: Evolution of the Late Ordovician orthid brachiopod Gnamptorhynchos Jin, 1989 from Platystrophia King, 1850, in North America. *Journal of Paleontology 74*, 983–991.

Jin, J. & Zhan, R.-B., 2001: *Late Ordovician Articulate brachiopods from the Red River and Stony Mountain Formations, Southern Manitoba.* NRC Research Press, Ottawa, Ontario, Canada, 117 pp.

Jin, J. & Zhan, R.-b. 2008: *Late Ordovician Orthide and Billingsellide Brachiopods from Anticosti Island, Eastern Canada: Diversity Change Through Mass Extinction,* NRC Research Press, Ottawa, 151 pp.

Jin, J., Zhan, R.-b. & Rong, J.-y. 2006: Taxonomic reassessment of two virgianid brachiopod genera from the Upper Ordovician and Lower Silurian of South China. *Journal of Paleontology 80*, 72–82.

Jin, J., Zhan, R.-b., Copper, P. & Caldwell, W.G.E. 2007: Epipunctae and phosphatized setae in Late Ordovician plaesiomyid brachiopods from Anticosti Island, Eastern Canada. *Journal of Paleontology 81*, 666–684.

Jones, O.T. 1928: Plectambonites and some allied genera. *Memoirs of the Geological Survey of Great Britain, Palaeontology 1*, 367–527, pls. 321–325.

Kanygin, A.V., Moskalenko, T.A., Yadrenkina, A.G., Abaimova, G.P., Semenova, B.C., Sychev, O.V. & Timokhin, A.V. 1989: *The Ordovician of the Siberian Platform. Fauna and Stratigraphy of the Lena Facies Zone.* Nauka, Novosibirsk, [in Russian], 216 pp.

Kanygin, A.V., Timokhin, A.V., Sychev, O.V. & Yadrenkina, A.G. 2006: Osnovnye etapy evolutsii i biofacial'noe raionirovanie ordovikskogo paleobasseina Sibirskoi platform [The main Ordovician evolutionary stages and biofacies of the of the Siberian platform basin], 22–25. *Field excursion guidebook. Contributions of the International Symposium 'Palaeogeography and global Correlation of Ordovician events,' Novosibirsk, August 5–16, 2006.* Academic Publishing House 'Geo', Novosibirsk, Russia, 90 pp.

King, W. 1846: Remarks on certain genera belonging to the class Palliobranchiata. *Annals and Magazine of Natural History (Series 1) 18*, 26–42, pls. 83–94.

King, W. 1850: A monograph of the Permian Fossils of England. *Palaeontological Society Monograph 3*, xxxvii + 258 pp.

Klenina, L.N., Nikitin, I.F. & Popov, L.E. 1984: *Brachiopods and biostratigraphy of Middle and Upper Ordovician of Khingiz Mountains.* Nauka, Kazakhstan SSR, Alma-Ata, 196 pp. [In Russian].

Kozłowski, R. 1929: Les brachiopodes gotlandiens de la Podolie polonaise. *Palaeontologia Polonica 1*, 1–254.

Kuhn, O. 1949: *Lehrbuch der Paläozoologie,* 326. E. Schweizerbart'sche Verlagsbuchhandlung, Stuttgart.

Kulkov, N.P. & Severgina, L.G. 1989: *Ordovician and Silurian Stratigraphy and Brachiopods from Gorny Altai,* Nauka, Moscow, 221 pp. [in Russian].

Lamont, A. 1935: The Drummuck Group, Girvan: stratigraphical revision with descriptions of new fossils from the lower part of the group. *Transactions of the Geological Society of Glasgow 19*, 288–334.

Laurie, J.R. 1991: Articulate brachiopods from the Ordovician and Lower Silurian of Tasmania. *Memoirs of the Association of Australasian Palaeontologists 11*, 1–106.

Leone, F., Hammann, W., Laske, R., Serpagli, E. & Villas, E. 1991: Lithostratigraphic units and biostratigraphy of the post-sardic Ordovician sequence in south-west Sardinia. *Bollettino della Società Geologica Italiana 30*, 201–235.

Linnaeus, C. 1758: *Systema Naturae.* Laurentii Salvii, Stockholm, 824 pp.

Liu, D.-y., Xu, H.-k. & Liang, W.-p. 1983: *Phylum Brachiopoda. Palaeontological Atlas of Eastern China. Early Palaeozoic Vol. 1.* Science Press, Beijing, 657 pp. 176 pls.

Lockley, M.G. & Williams, A. 1981: Lower Ordovician Brachiopoda from mid and southwest Wales. *Bulletin of the British Museum (Natural History), Geology Series 35*, 1–78.

Ludvigsen, R. 1975: Ordovician formations and faunas, southern Mackenzie Mountains. *Canadian Journal of Earth Sciences 12*, 663–697.

Macomber, R.W. 1970: Articulate brachiopods from the Upper Bighorn Formation (Late Ordovician) of Wyoming. *Journal of Paleontology 44*, 416–450.

McChesney, J.H. 1861: *Descriptions of New Fossils from the Paleozoic Rocks of the Western States.* Transactions of the Chicago Academy of Sciences, 77–95.

McCoy, F. 1844: A synopsis of the characters of the Carboniferous Limestone fossils of Ireland. Dublin. 207 pp.

McCoy, F. 1852: *Description of the British Paleozoic Fossils in the Geological Museum of the University of Cambridge.* J. W. Parker & Son, London, 185–644.

McKerrow, W.S. & Cocks, L.R.M. 1981: Stratigraphy of the eastern Bay of Exploits, Newfoundland. *Canadian Journal of Earth Sciences 18*, 751–764.

Measures, E.A., Rohr, D.M. & Blodgett, R.B. 1992: Depositional environments and some aspects of the fauna of Middle Ordovician rocks of the Telsitna Formation, Northern Kuskokwim Mountains, Alaska. *U.S. Geological Survey Bulletin 2041*, 186–196.

Mélou, M. 1990: Brachiopodes articules de la Coupe de l'Ile de Rosan (Crozon, Finistére). Formation des Tufs et Calcaires de Rosan (Caradoc–Ashgill). *Géobios 23*, 539–579.

Menakova, G.N. 1984: O nekotorykh brakhiopodykh verkhnego ordovika Zeravshano-Gissarskoi Gornoi Oblasti [On certain brachiopods of the Upper Ordovician of Zeravshan–Gissar Mountain region]. *In* Dzhalilov, M.R. (ed.). *Novye Vidy Iskopaemoi Flory i Fauny Tadzhikista,* 75–85. Institut Geologii Donish, Dushanbe, Akademiia Nauk Tadzhikskoi SSR., [In Russian].

Mergl, M. 1990: Late Ordovician Foliomena brachiopod fauna from the Holy Cross Mountains (Poland). *Casopis Pro Mineralogii Geologii 35*, 147–154.

Mergl, M. 2006: A review of Silurian discinoid brachiopods from historical British localities. *Bulletin of Geosciences 81*, 215–236.

Miall, M.A. 2008: The Paleozoic western craton margin, 181–211. *In* Miall, A.D. (ed.): *Sedimentary basins of the World, vol. 5: The sedimentary basins of the United States and Canada.* Elsevier B.V., Xiii + 593 pp.

Mitchell, W.I. 1977: The Ordovician brachiopoda from Pomeroy, Co. *Tyrone. Palaeontological Society Monograph 130*, 1–138.

Moore, R.C. 1952: Brachiopoda. 197–267. *In* Moore, R.C., Lalicker, C.G. & Fischer, A.G. (eds): *Invertebrate Fossils.* McGraw-Hill, New York, 766 pp.

Neuman, R.B. 1994: Late Ordovician (Ashgill) *Foliomena* fauna brachiopods from northeastern Maine. *Journal of Paleontology 68*, 1218–1234.

Neuman, R.B., Bruton, D.L. & Pojeta, J. 1997: Fossils from the Ordovician 'Upper Hovin Group' (Caradoc–Ashgill), Trondheim region, Norway. *Norges Geologiske Undersøkelse, Bulletin 432*, 25–58.

Nielsen, A.T. 2004: Ordovician Sea Level Changes: a Baltoscandian Perspective. 84–93. *In* Webby, B.D., Paris, F., Droser, M.L. & Percival, I.C. (eds): *The Great Ordovician Biodiversification Event.* Columbia University Press, New York, 484 pp.

Nikiforova, O.I. 1978: Brachiopods of the Chashmankolon, Archalyk and Minkuchar beds. *Akademiia Nauk SSSR, Sibirskoe Otdelenie, Institut Geologii i Geofiziki, Trudy 397*, 102–126, pls. 18–23.

Nikiforova, O.I. & Andreeva, O.A. 1961: Ordovician and Silurian stratigraphy of the Siberian Platform and its paleontological basis (Brachiopoda). *VSEGEI New Series 56*, 425 pp.

Nikiforova, O.I. & Sapelnikov, V.P. 1973: Some fossil pentamerides from the Zeravshan Range. *Institut geologii i geohimii, Ural'skij nauènyj centr, Akademiâ nauk SSSR, Trudy 99*, 89–114. [in Russian].

Nikitin, I.F. & Popov, L.E. 1984: Brakhiopody bestamakskoy i sargaldakskoy svit (sredniy ordovik). 126–166. *In* Klenina, L.N., Nikitin, I.F. & Popov, L.E. (eds): *Brakhiopody i biostratigrafiya srednego i verkhnego Ordovika Khrebta Chingiz.* Akademiya Nauka Kazakhskoy SSR, Alma-Ata, 196 pp. [in Russian].

Nikitin, I.F. & Popov, L.E. 1996: Strophomenid and triplesiid brachiopods from an Upper Ordovician carbonate mound in central Kazakhstan. *Alcheringa 20*, 1–20.

Nikitin, I.F., Popov, L.E. & Bassett, M.G. 2003: Late Ordovician brachiopods from the Selety river basin, north Central Kazakhstan. *Acta palaeontologica polonica 48*, 39–54.

Nikitin, I.F., Popov, L.E. & Bassett, M.G., 2006: Late Ordovician rhynchonelliformean brachiopods of north-central Kazakhstan, 223 -294. In Bassett M.G., Deisler V.K. (eds): *Studies in Palaeozoic Palaeontology: National Museum of Wales Geological Series, 25.* 294 pp.

Nikolaev, A.A. & Sapelnikov, V.P. 1969: Dva novykh roda pozdneordovikskikh Virgianidae [Two new genera of Late Ordovician Virgianidae]. *Trudy Sverdlovskogo Ordena Trudovogo Krasnogo Znameni Gornogo Instituta 63*, 11–17. [in Russian].

Öpik, A. 1930: Brachiopoda protremata der Estländischen Ordovizischen Kukruse-stufe. *Universitatis Tartuensis (Dorpatensis). Serie A mathematica, physica, medica, XVII*, 261 pp. Tartu.

Öpik, A. 1933: Über Plectamboniten. *Universitatis Tartuensis (Dorpatensis). Acta et Commentationes (series A) 24*, 1–79, pls. 1–12.

Öpik, A. 1934: Über Klitamboniten. *Acta et commentationes.Universitatis Tartuensis (Dorpatensis). Serie A mathematica, physica, medica XXVI*, 1–239.

Oradovskaya, M.M. 1983: [Brachiopods]. 35–73. *In* Koren, T.N., Oradovskaya, M.M., Pylma, L.J., Sobolevskaya, R.F. & Chugaeva, M.N. (eds): *The Ordovician and Silurian Boundary in the Northeast of the USSR.* Leningrad, Nauka, 205 pp. [in Russian].

Oradovskaya, M.M. 1988: The Ordovician System in north-east and far east of the USSR, 85–115. *In* Ross Jr., R.J & Talent, J. (eds): *The Ordovician System in most of Russian Asia.* International Union of Geological Sciences (IUGS) *26*. 115 pp.

Parkes, M.A. 1992: Caradoc brachiopods from the Leinster terrane (SE Ireland) - a lost piece of the Iapetus puzzle? *Terra Nova 4*, 223–230.

Parkes, M.A. 1994: The brachiopods of the Duncannon Group (Middle-Upper Ordovician) of southeast Ireland, Bulletin of the British Museum (Natural History), *Geology Series 50*, 105–174.

Paškevičius, J. 1994: *Baltijos Respubliky Geologija [The Geology of the Baltic Republics].* Vastybins leidybos centras, Vilnius, 447 pp.

Patzkowsky, M.E. 1995: Gradient analysis of middle Ordovician brachiopod biofacies: biostratigraphic, biogeographic, and macroevolutionary implications. *Palaios 10*, 154–179.

Patzkowsky, M.E. & Holland, S.M. 1997: Patterns of Turnover in Middle and Upper Ordovician Brachiopods of the Eastern United States: a test of Coordinated Stasis. *Paleobiology 23*, 420–443.

Percival, I.G. 1991: Late Ordovician articulate brachiopods from central New South Wales. *Memoirs of the Association of Australasian Palaeontologists 11*, 107–177.

Percival, I.G. 2009: Late Ordovician strophomenide and pentameride brachiopods from central New South Wales. *Proceedings of the Linnean Society of New South Wales, 130*, 157–178.

Popov, L.E. 1980: Novye strofomenidy srednego ordovika severnogo Kazakhstan [New strophomenids from the Middle Ordovician of Northern Kazatchstan], 54–57. *In* Stukalina, G.A. (ed.): *Novye vidy drevnikh rasteniy i bespozvonochnykh SSSR*, Nauka, Moskva. [in Russian].

Popov, L.E. 1985: Brachiopods of the Anderken horizon of the Chu-Ili Hills (Kazakhstan). *Ezhegodnik soyuznovo Palaeontologicheskovo Obshchestva 28*, 50–68.

Popov, L.E. & Cocks, L.R.M. 2006: Late Ordovician brachiopods from the Dulankara Formation of the Chu-Ili Range, Kazakhstan: their systematics, palaeoecology and palaeobiogeography. *Palaeontology 49*, 247–283.

Popov, L.E., Nikitin, I.N. & Cocks, L.R.M. 2000: Late Ordovician brachiopods from the Otar Member of the Chu-Ili Range, South Kazakhstan. *Palaeontology 43*, 833–870.

Popov, L.E., Cocks, L.R.M. & Nikitin, I.F. 2002: Upper Ordovician brachiopods from the Formation, Kazakhstan: their ecology and systematics. *Bulletin of the British Museum (Natural History), Geology Series 58*, 13–79.

Potter, A.W. 1984: Palaeobiogeographical relations of Late Ordovician brachiopods from the York and Nixon Folk Terranes, Alaska. *Geological Society of America, Abstracts with Programs 16*, p.626.

Potter, A.W. 1990a: Middle and Late Ordovician brachiopods from the Eastern Klamath Mountains, Northern California, Part 1. *Palaeontographica A 212*, 31–158, pls. 151–157.

Potter, A.W. 1990b: Middle and Late Ordovician brachiopods from the Eastern Klamath Mountains, Northern California, Part 2. *Palaeontographica A 213*, 1–114, pls. 1–110.

Potter, A.W. 1990c: The Ordovician brachiopod genus *Bimuria* from the Eastern Klamath Mountains, Northern California. *Journal of Paleontology 64*, 200–213.

Potter, A.W. 1991: Discussion of the systematic placement of the Ordovician brachiopod genera Cooperea and Craspedelia by Cocks and Rong (1989). *Journal of Paleontology 65*, 742–755.

Potter, A.W. & Blodgett, R.B. 1992: Paleobiogeographic relations of Ordovician brachiopods from the Nixon Folk Terrane, West Central Alaska. *Geological Society of America, Abstracts with Programs 24*, p.76.

Potter, A.W. & Boucot, A.J. 1971: Ashgillian, Late Ordovician brachiopods from the eastern Klamath Mountains of northern California. *Geological Society of America, Abstracts with Programs 3*, 180–181.

Potter, A.W. & Boucot, A.J. 1992: Middle and Late Ordovician brachiopod benthic assemblages of North America. 307–327. *In* Webby, B.D. & Laurie, J.R. (eds): *Global Perspectives on Ordovician Geology.* Balkema, Rotterdam, 513 pp.

Potter, A.W., Gilbert, W.G., Ormiston, A.R. & Blodgett, R.B. 1980: Middle and Upper Ordovician brachiopods from Alaska and northern California and the paleogeographic implications. *Geological Society of America, Abstracts with Programs 12*, p.147.

Potter, A.W., Blodgett, R.B. & Rohr, D.M. 1988: Paleogeographic relations and paleogeographic significance of Late Ordovician brachiopods from Alaska. *Proceedings of the Geological Society of America, Abstracts with Programs, 20*, p.339.

Price, D. 1981: Ashgill trilobite faunas from the Llyn Peninsula, North Wales. *Geological Journal 16*, 201–216.

Rasmussen, C.M.Ø. 2009: A palaeoecological and biogeographical investigation of the Ordovician–Silurian boundary based on brachiopods. Unpublished PhD Thesis, University of Copenhagen, Copenhagen, 82 pp., 8 pls.

Rasmussen, C.M.Ø. 2011: Final destination, first discovered: the tale of *Oanduporella* Hints, 1975. *In* Gutiérrez-Marco, J.C, Rabano, I. & Garcia-Bellido, D. (eds): *Ordovician of the World*, 447–455. Cuadernos del Museo Geominero, 14. Instituto Geológico y Minero de España, Madrid.

Rasmussen, C.M.Ø. & Harper, D.A.T. 2010: An atypical Late Ordovician Red River brachiopod fauna from eastern North Greenland: Eastward, offshore migration of the Hiscobeccus-fauna. Abstracts, Annual Meeting of the Palaeontological Association, University of Ghent, 17th–20th December 2010, p. 68.

Rasmussen, C.M.Ø. & Harper, D.A.T. 2011a: Interrogation of distributional data for the end Ordovician crisis interval: where did disaster strike? *Geological Journal 46*, 478–500.

Rasmussen, C.M.Ø. & Harper, D.A.T. 2011b: Did the amalgamation of continents drive the End Ordovician mass extinctions? *Palaeogeography, Palaeoclimatology, Palaeoecology 331*, 48–62.

Rasmussen, C.M.Ø., Hansen, J. & Harper, D.A.T. 2007: Baltica: a mid Ordovician diversity hotspot. *Historical Biology 19*, 255–161.

Rasmussen, C.M.Ø., Ebbestad, J.O.R. & Harper, D.A.T. 2010: Unravelling a Late Ordovician pentameride (Brachiopoda) hotspot from the Boda Limestone, Siljan district, central Sweden. *GFF 132*, 133–152.

Reed, F.R.C. 1917: The Ordovician and Silurian brachiopoda of the Girvan district. *Transactions of the Royal Society of Edinburgh: Earth Sciences 51*, 795–998, pls. 1–24.

Reed, F.R.C. 1932: *Report on the Brachiopods from the Trondheim area [Norway], volume 4*, 115–146. Skrifter utgitt av Det Norske Videnskaps-Akademi: Matematisk-Naturvidenskapelig Klasse, Oslo, pls. 118–122.

Rigby, J.K., Potter, A.W. & Blodgett, R.B. 1988: Ordovician Sphinctozoan sponges of Alaska and Yukon Territory. *Journal of Paleontology 62*, 731–746.

Rigby, J.K., Karl, S.M., Blodgett, R.B. & Baichtal, J.F. 2005: Ordovician 'sphinctozoan' sponges from Prince of Wales Island, southeastern Alaska. *Journal of Paleontology 79*, 862–870.

Rigby, J.K., Blodgett, R.B. & Britt, B.B. 2008: Ordovician sponges from west-central and east central Alaska and western Yukon Territory, Canada. *Bulletin of Geosciences, 83*, 153–168.

Rigby, J.K., Blodgett, R.B. & Anderson, N.K. 2009: Emsian (late Early Devonian) sponges from west.central and south-central Alaska. *Journal of Paleontology 83*, 293–298.

Rohr, D.M. & Blodgett, R.B. 1985: Upper Ordovician gastropoda from West-Central Alaska. *Journal of Paleontology 59*, 667–673.

Rohr, D.M., Dutro, Jr. J.T. & Blodgett, R.B. 1992: Gastropods and brachiopods from the Ordovician Telsitna Formation, northern Kuskokwim Mountains, west-central Alaska, 499–512. *In* Webby, B.D. & Laurie, J.R. (eds): *Global Perspectives on Ordovician Geology*. Balkema, Rotterdam, 513 pp.

Rong, J.-y. 1979: The *Hirnantia* fauna of China with the comments on the Ordovician – Silurian boundary. *Journal of Stratigraphy (formerly Acta Stratigraphica Sinica) 3*, 1–29.

Rong, J.-y. 1984a: Brachiopods of the latest Ordovician in the Yichang district, western Hubei, 111–176. *In* Nanjing Institute of Geology and Palaeontology, Academia Sinica (ed.): *Stratigraphy and Palaeontology of systematic boundaries in China*. Ordovician–Silurian boundary 1. Anhui Science and Technology Publishing House, Hefei, China, 516 pp.

Rong, J.-y. 1984b: Distribution of the *Hirnantia* fauna and its meaning, 101–112. *In* Bruton, D.L. (ed.): *Aspects of the Ordovician System*. Universitetsforlaget, Oslo, 228 pp.

Rong, J.-y. & Boucot, A.J. 1998: A global review of the Virgianidae (Ashgillian – Llandovery, Brachiopoda, Pentameroidea). *Journal of Paleontology 72*, 457–465.

Rong, J.-y. & Cocks, L.R.M. 1994: True *Strophomena* and a revision of the classification and evolution of strophomenoid and 'strophodontoid' brachiopods. *Palaeontology 37*, 651–694.

Rong, J.-y. & Zhan, R.-b. 1996: Brachidia of Late Ordovician and Silurian eospiriferines (Brachiopoda) and the origin of the spiriferides. *Palaeontology 39*, 941–977.

Rong, J.-y., Zhan, R.-b. & Han, N.-r. 1994: The oldest known *Eospirifer* (Brachiopoda) in the Changwu Formation (late Ordovician) of western Zhejiang, east China, with a review of the earliest spiriferoids. *Journal of Paleontology 68*, 763–776.

Rong, J.-y., Zhan, R.-b. & Harper, D.A.T. 1999: Late Ordovician (Caradoc–Ashgill) brachiopod faunas with *Foliomena* based on data from China. *Palaios 14*, 412–431.

Rong, J.-y., Chen, X. & Harper, D.A.T. 2002: The latest Ordovician Hirnantia Fauna (Brachiopoda) in time and space. *Lethaia 35*, 231–249.

Rong, J.-y., Jin, J. & Zhan, R.-b. 2007: Early Silurian Sulcipentamerus and related pentamerid brachiopods from South China. *Palaeontology 50*, 245–266.

Rõõmusoks, A. 1959: Strophomenoidea of the Ordovician and Silurian of Estonia, I. The genus *Sowerbyella* Jones. *Tartu Riikliku Ulikool Toimetised 75*, 11–50.

Rõõmusoks, A. 1963: Strophomenoidea of the Ordovician and Silurian of Estonia: II New genera and species from the Harju Series. *Eesti NSV Teaduste Akadeemia, Keemia Bioloogia 12*, 231–241, pls. 231–232.

Rõõmusoks, A. 1964: *Some brachiopods from the Ordovician of Estonia*. Tartu Ülikooli Toimetised, *153*, 3–29, [in Russian].

Rõõmusoks, A. 1989: Über die divergenz der Leptaenidae (brachiopoda) in der Viru- und Harju- zeit in Baltoskandia. *Eesti NSV Teaduste Akadeemia Toimetised, Geoloogia 38*, 112–117.

Rõõmusoks, A. 2004: Ordovician strophomenoid brachiopods of northern Estonia. *Fossilia Baltica 3*, 1–151.

Ross, J.R. & Dutro, J.T. 1966: Silicified Ordovician brachiopods from East-Central Alaska. *Smithsonian Miscellaneous Collections 149*, 22 pp.

Rozman, K.S. 1964: Brakiopody srednogo i poznego ordovika Selenyakhskogo kryazha [Middle and Late Ordovician brachiopods from the Selennayakh district], 109–193. *In* Chugaeva, M.N., Rozman, K.S. & Ivanova, V.A. (eds): *Stravnitelnaya biostratigrafiya ordovikskikh otlozheniy Severo-Vostoka SSSR [Comparative biostratigraphy of Ordovician deposits in the north-east of the USSR], volume 106*. Akadamia Nauka SSSR Geologicheskii Institut, Trudy, 233 pp [in Russian].

Rozman, K.S. 1978: Brachiopods of the Obikalon Beds. Boundary Beds of the Ordovician and Silurian of the Altae-Sayan Region and Tien-Shan. *Transactions of the Academy of USSR Siberian Branch Institute of Geology and Geophysics 397*, 75–125, [in Russian].

Rzhonsnitskaia, M.A. 1960: Order Atrypida, 257–264. *In* Sarytcheva, T.G. & Orlov, Y.A. (eds): *Mshanki, Brakhiopody [Bryozoa, Brachiopoda]. Osnovy Paleontologii 7*. Akademia Nauk SSSR, Moscow. [in Russian].

Sainsbury, C.L. 1965: Previously undescribed Middle (?) Ordovician, Devonian (?) and Cretaceous (?) rocks, White Mountain area, near McGrath, Alaska. *United States Geological Survey, Professional Paper 525-C*, C91–C95.

Sapelnikov, V.P. 1985: *Sistema i stratigraficheskoye znacheniye brakhiopod podotryada pentameridin [System and Stratigraphic Significance of the Suborder Pentameridina Brachiopods]*. Instituta geologii i Geokhimii, Ural'skii Nauchniyi Tsentr, Akademiya Nauk SSSR, Moscow, Nauka, 206 pp [In Russian].

Sapelnikov, V.P. & Rukavishnikova, T.B. 1975: *Verkhneordovikskiye, Siluriyskiye i Nizhnedevonskiye pentameridy Kazakhstana [The Upper Ordovician, Silurian, and Lower Devonian pentameroids of Kazakhstan]*. Akademiia Nauk SSSR, Uralskiy Nauchnyy Tsentr, Moskva, Nauka, 227 pp [In Russian].

Sapelnikov, V.P. & Rukavishnikova, T.B. 1976: *Galeatellina* novoe nazvanie roda pentamerid (Brachiopoda) [*Galeatellina*, a new name for a pentamerid genus (Brachiopoda)]. *Paleontologicheskii Zhurnal 2*, 122 pp [In Russian].

Sardeson, F.W. 1892: The range and distribution of the lower Silurian faunas of Minnesota with descriptions of some new species. *Bulletin of the Minnesota Academy of Natural Science 3*, 326–343.

Savage, T.E. 1913: Alexandrian series in Missouri and Illinois. *Geological Society of America, Bulletin 24*, 351–376.

Schuchert, C. 1893: Classification of the Brachiopoda. *American Geologist 11*, 141–167.

Schuchert, C. 1913: Class 2. Brachiopoda, 355–420. *In* von Zittel, K.A. (ed.): *Text-book of Palaeontology, volume 1*, part 1, 2nd edition, translated and edited by C. R. Eastman. MacMillan & Co. Ltd., London, 839 pp.

Schuchert, C. & Cooper, G.A. 1930: Upper Ordovician and Lower Devonian stratigraphy and paleontology of Percé, Quebec: part 2, New species from the Upper Ordovician of Percé. *American Journal of Science (Series 5) 20*, 265–288.

Schuchert, C. & Cooper, G.A. 1931: Synopsis of the brachiopod genera Orthoidea and Pentameroidea, with notes on the Telotremata. *American Journal of Science (Series 5) 22*, 241–255.

Schuchert, C. & Cooper, G.A. 1932: Brachiopod genera of the suborders Orthoidea and Pentameroidea. *Memoirs of the Peabody Museum of Natural History 4*, xii + 270 pp.

Schuchert, C. & LeVene, C.M. 1929: Brachiopoda (generum et genotyporum index et bibliographia). *In* Pompeckj, J.F. (ed.): *Fossilium Catalogus 1, Animalia, Pars 42*. W. Junk, Berlin, 140 pp.

Sennikov, N.V., Yolkin, E.A., Petrunina, Z.E., Gladkikh, L.A., Obut, O.T., Izokh, N.G. & Kipriyanova, T.P. 2008: Ordovician-Silurian biostratigraphy and paleogeography of the Gorny Altay. in Sennikov, N.V., Kanygin, A.V. (eds), Publishing House Sb Ras, Novosibirisk, pp. 1–156.

Severgina, L.G. 1978: Brachiopods and stratigraphy og the Upper Ordovician of mountainous Altai Region, Salair and mountainous Shoriia. *Akademiia Nauk SSSR, Sibirskoe Otdelenie, Institut Geologii i Geofiziki, Trudy, 405*, 3–41. [in Russian].

Severgina, L.G. 1984: Some Upper Ordovician (Ashgill) brachiopods of the Gornoi Altai. *Trudy Instituta Geologii i Geofiziki Akademia Nauk SSSR, Sibirskoe Ortdelenie, Moscow 584*, 29–48, pls. 3–4.

Sheehan, P. M. 1973: The relation of Late Ordovician glaciation to the Ordovician – Silurianm changeover in North American brachiopod faunas. *Lethaia 6*, 147–154.

Sheehan, P.M. 1979: Swedish Late Ordovician marine benthic assemblages and their bearing on brachiopod zoogeography, 61–73. *In* Gray, J. & Boucot, A.J. (eds): *Historical Biogeography, Plate Tectonics, and the Changing Environment*, xii + 500 pp. The Oregon State University Press, Corvallis.

Sheehan, P.M. 1987: Late Ordovician (Ashgillian) brachiopods from the region of the Sambre and Muese rivers, Belgium. *Bulletin de l'Institut Royal des Sciences Naturelles de Belgique, Sciences de la Terre 57*, 5–81.

Sheehan, P.M. & Lespérance, P.J. 1979: Late Ordovician (Ashgillian) brachiopods from the Percé region of Québec. *Journal of Paleontology 53*, 950–967.

Sowerby, J.d.C. 1839: Shells, 579–712, pls. 37. *In* Murchison, R.I. (ed.): *The Silurian System*. John Murray, London, 768 pp.

Spjeldnæs, N.N. 1957: The Middle Ordovician of the Oslo Region, Norway. 8, Brachiopods of the Suborder Strophomenida. *Norsk Geologisk Tidsskrift 37*, 1–214.

Suárez-Soruco, R. 1992: El Paleozoico inferior de Bolivia y Perú, 225–239. *In* Gutierrez-Marco, J.C., Saavedra, J. & Rábano, I. (eds): *Paleozoico Inferior de Iberomérica*. Universida de Extremadura, Mérida, Spain.

Teichert, C. 1937: *Ordovician and Silurian faunas from Arctic Canada. Report on the 5th Thule Expedition 1921–1924, volume 1, no. 5.* Gyldendal, Copenhagen, 169pp.

Torsvik, T.H. & Cocks, L.R.M. 2009: The Lower Palaeozoic palaeogeographical evlution of the northeastern and eastern peri-Gondwanan margin from Turkey to New Zealand, 3–21. *In* Bassett, M.G. (ed.): *Early Palaeozoic Peri-Gondwana Terranes: new insights from tectonics and biogeography*. The Geological Society, London, Special Publications, *325*, 288pp.

Torsvik, T.H., Smethurst, M.A., Voo, R.V.D., Trench, A., Abrahamsen, N. & Halvorsen, E. 1992: Baltica. A synopsis of Vendian – Permian palaeomagnetic data and their palaeotectonic implications. *Earth-Science Reviews 33*, 133–152.

Torsvik, T.H., Smethurst, M.A., Meert, J.G., Van der Voo, R., McKerrow, W.S., Brasier, M.D., Stort, B.A. & Walderhaug, H.J. 1996: Continental break-up in the Neoprotorozoic and Palaeozoic – A tale of Baltica and Laurentia. *Earth-Science Reviews 40*, 229–258.

Ulrich, E.O. & Cooper, G.A. 1936: New genera and species of Ozarkian and Canadian brachiopods. *Journal of Paleontology 10*, 616–631.

Ulrich, E.O. & Cooper, G.A. 1942: New genera of Ordovician brachiopods. *Journal of Paleontology 16*, 620–626.

Villas, P.E. 1985: Braquiópodos del Ordovícico Medio y Superior de las Cadenas Ibericas Orientales. *Memorias del Museo Paleontólogico de la Universidad de Zaragoza 1*, 223 pp.

Villas, E. 1992: New Caradoc brachiopods from the Iberian Chains (northeastern Spain) and their stratigraphic significance. *Journal of Paleontology 66*, 772–793.

Villas, E., Hammann, W. & Harper, D.A.T. 2002: *Foliomena* fauna (brachiopoda) from the Upper Ordovician of Sardinia. *Palaeontology 45*, 267–295.

Waagen, W. 1884: Salt Range fossils, vol I, part 4. *Memoirs of the Geological Survey of India, Palaeontologia Indica (series 13) 3–4*, 547–728.

Waagen, W. 1885: Salt Range fossils, vol. I, part 4. Productus Limestone fossils, Brachiopoda. *Memoirs of the Geological Survey of India, Palaeontologia Indica (series 13) 5*, 729–770.

Waisfeld, B.G., Sánchez, T.M., Benedetto, J.L. & Carrera, M.G. 2003: Early Ordovician (Arenig) faunal assemblages from western Argentina: biodiversification trends in different geodynamic and palaeogeographic settings. *Palaeogeography, Palaeoclimatology, palaeoecology 196*, 343–373.

Wang, Y. 1949: Maquoketa Blachiopoda of Iowa. *Geological Society of America Memoir 42*, 1–55.

Willard, B. 1928: The brachiopods of the Ottosee and Holston formations of Tennessee and Virginia. *Bulletin of Harvard Museum of Comparative Zoology 68*, 255–292.

Williams, A. 1953: The classification of the strophomenoid brachiopods. *Washington Academy of Sciences Journal 43*, 1–13.

Williams, A. 1962: The Barr and Lower Ardmillan Series (Caradoc) of the Girvan District, southwest Ayrshire. *Geological Society of London Memoir 3*, 1–267, pls. 1–25.

Williams, A. 1963: The Caradocian brachiopod faunas of the Bala district, Merionetshire. *Bulletin of the British Museum (Natural History), Geology Series 8*, 327–341, pls. 1–16.

Williams, A. 1965: Suborder Strophomenidina. 362–412. *In* Moore, R.C. (ed.): *Treatise on Invertebrate Paleontology. Part H, Brachiopoda, 1*. The Geological Society of America and the University of Kansas Press, Lawrence, Kansas, 521 pp.

Williams, A. & Curry, G.B. 1985: Lower Ordovician Brachiopoda from the Tourmakeady Limestone, Co. *Mayo, Ireland. Bulletin of the British Museum (Natural History), Geology Series 38*, 183–269.

Williams, A. & Harper, D.A.T. 2000: Order Orthida, 714–846. *In* Kaesler, R.L. (ed.): *Treatise on invertebrate palaeontology: Part H, revised. Brachiopoda, 3*. Geological Society of America, and The University of Kansas Press, Boulder, Colorado & Lawrence, Kansas, 919 pp.

Williams, A. & Wright, A.D. 1981: The Ordovician – Silurian boundary in the Garth area of southwest Powys, Wales. *Geological Journal 16*, 1–39.

Williams, A., Carlson, S.J., Brunton, C.H.C., Holmer, L.E. & Popov, L.E. 1996: A supra- ordinal classification of the Brachiopoda. *Philosophical Transactions of the Royal Society of London (Series B) 351*, 1171–1193.

Winchell, N.H. & Schuchert, C. 1893: The lower Silurian Brachiopoda of Minnesota, 333–474. *In* Lesquereux, L., Woodward, A., Thomas, B.W., Schuchert, C., Ulrich, E.O. & Winchell, N.H. (eds): *The Geology of Minnesota, 3(1)*. Final Report Harrison & Smith, Geological and Natural History Survey of Minnesota, Minneapolis.

Winchell, N.H. & Schuchert, C. 1895: The Lower Silurian Brachiopoda of Minnesota. *Minnesota Geological Survey, Final Report, Paleontology 3*, 333–474.

Woodward, S.P. 1852: *A manual of the Mollusca; or rudimentary treatise of recent and fossil shells*. John Weale, London, xvi + 486 pp.

Wright, A.D. 1964: The fauna of the Portrane Limestone, II. *Bulletin of the British Museum (Natural History), Geology 9*, 157–256.

Wright, A.D. 1968: A new genus of dicoelosiid brachiopod from Dalarna. *Arkiv för Zoologi 22*, 127–138.

Wright, A.D. 1982: A new species of *Ptychopleurella* (Brachiopoda) from the Upper Ordovician Boda Limestone of Dalarna, Sweden. *Journal of Paleontology 56*, 351–357.

Wright, A.D. & Jaanusson, V. 1993: New genera of Upper Ordovician triplesiid brachiopods from Sweden. *GFF 115*, 93–108.

Xu, H.-k. 1996: Late Ordovician brachiopods from the central part of eastern Qinling Region. *Acta palaeontologica Sinica 35*, 544–574. [in Chinese, with English summary].

Xu, H.-k., Rong, J.-y. & Liu, D.-y. 1974: Ordovician brachiopods. *In* Nanjing Institute of Geology and Palaeontology, Academia Sinica (ed.): *Handbook of Stratigraphy and Palaeontology in Southwest China, Beijing*, 144–154. pls 164–166.

Zhan, R.-b. & Cocks, L.R.M. 1998: Late Ordovician brachiopods from the South China Plate and their palaeogeographical significance. *Special Papers in Palaeontology 59*, 1–70.

Zhan, R.-b. & Harper, D.A.T. 2006: Biotic diachroneity during the Ordovician Radiation: evidence from South China. *Lethaia 39*, 211–226.

Zhan, R.-b. & Jin, J. 2005: New data on the *Foliomena* fauna (brachiopoda) from the Upper Ordovician of South China. *Journal of Paleontology 79*, 670–686.

Zhan, R.-b., Rong, J.-y., Jin, J. & Cocks, L.R.M. 2002: Late Ordovician brachiopod communities of southeast China. *Canadian Journal of Earth Sciences 39*, 445–468.

Zhang, N. & Boucot, A.J. 1988: *Epitomyonia* (brachiopoda): ecology and functional morphology. *Journal of Paleontology 62*, 753–758.

Zhang, S. & Barnes, C.R. 2007: Late Ordovician to early Silurian conodont faunas from the Kolyma Terrane, Omulev Mountains, Northeast Russia, and their paleobiogeographic affinity. *Journal of Paleontology, 81*, 490–512.

Plates 1–17

Plate 1

1.   ***Acrosaccus* cf. *shuleri* Willard, 1928**
1:   Ventral exterior view (MGUH 29442). Locality A-1230.

2–8.   *Tetraphalerella planobesa* (**Cooper, 1956**) new combination
2:   View of ventral valve exterior (MGUH 29443). Locality A-1230.
3:   View of dorsal valve exterior (MGUH 29444). Locality A-1230.
4:   Interior view of ventral valve (MGUH 29445). Locality A-1230.
5–7:   Perpendicular and anterior views of dorsal cardinalia. Note the undercut bifid cardinal process and the arrangement of the brachiophores (MGUH 29446, MGUH 29447). Locality A-1230.
8:   Close-up of valve exterior showing punctate ornament (MGUH 29443).

9–13.   ***Transridgia costata* n. gen. et n. sp.**
9:   Exterior view of ventral valve (MGUH 29448). Locality A-1230.
10:   Exterior view of dorsal valve (MGUH 29449). Locality A-1230.
11:   Interior view of ventral valve (MGUH 29450). Locality A-1230.
12:   Interior view of dorsal valve (MGUH 29451, holotype). Locality A-1230.
13:   Anterior view of dorsal interior (MGUH 29449).

14–19.   *Leptaena* (*Leptaena*) sp.
14:   Ventral view of conjoined specimen (MGUH 29452). Locality A-1230.
15:   Posterior view of conjoined specimen showing ventral exterior rugation. Note also ventral interarea and delthyrial cavity (same specimen as in 14).
16:   Exterior view of dorsal valve (MGUH 29453). Locality A-1230.
17:   Posterolateral view of dorsal valve exterior. Note the sharp geniculation (same specimen as in 16).
18, 19:   Interior and oblique views of dorsal valve showing cardinalia with the widely spaced brachiophores. Note also the short median septum (MGUH 29454). Locality A-1230.

20–23.   ***Leptaena* (*Septomena*) *alaskensis* n. sp.**
20:   Fragment of ventral valve showing convexity of valve and possibly alate cardinal extremities (MGUH 29455). Locality A-1230.
21:   Exterior view of dorsal valve. Note ornament and geniculation along the commissure (MGUH 29456, holotype). Locality A-1230.
22:   Interior view of the holotype (MGUH 29456).
23:   Interior view of dorsal valve. Note the two side septa (MGUH 29457). Locality A-1230.

24.   ***Christiania aseptata* n. sp.**
24:   View of ventral valve exterior (MGUH X29458). Locality A-1230.

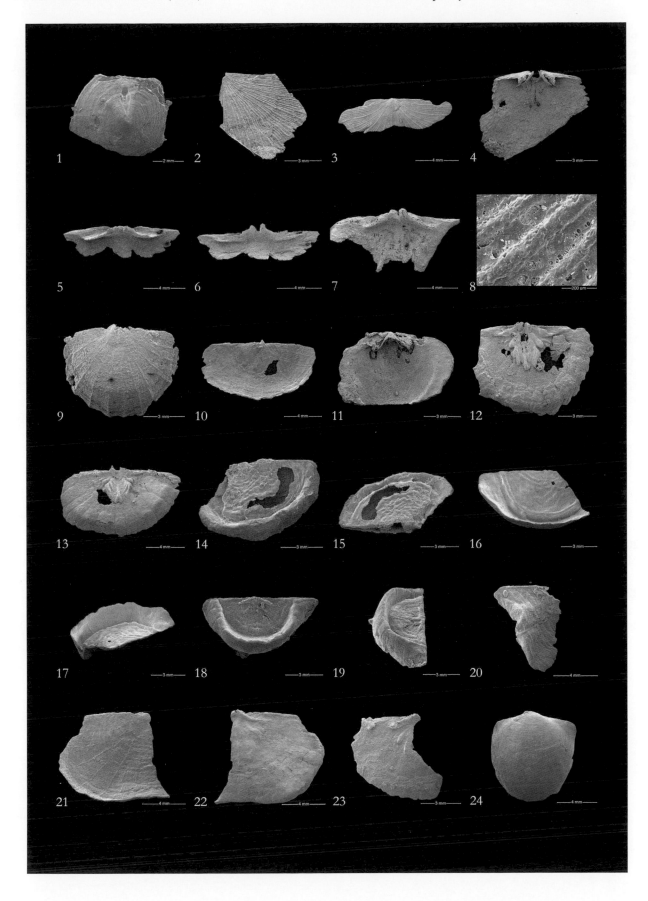

Plate 2

1–5. **Christiania aseptata n. sp.** (cont.)
1, 2: Dorsal valve exterior and posterior view of same specimen showing dorsal interarea (MGUH 29459). Locality A-1230.
3, 4: Perpendicular and anterior views of ventral valve interior (MGUH 29460). Locality A-1230.
5: Dorsal valve interior. Note the absence of a median septum (MGUH 29461, holotype). Locality A-1230.

6–7. **Strophomenoidea gen. et sp. indet.**
6: View of possibly ventral? exterior (MGUH 29462). Locality A-1230.
7: View of dorsal interior (MGUH 29463). Locality A-1230.

8–10. **Strophomenoidea gen. et sp. indet. 2**
8: View of ventral valve exterior. Note the juvenile *P. uniplicata* encrusting the valve (MGUH 29464). Locality A-1230.
9: View of ventral valve interior (MGUH 29464).
10: Fragment possibly showing dorsal valve exterior (MGUH 29465). Locality A-1230.

11–12. **Strophomenoidea gen. et sp. indet. 3**
11: Fragment of ventral exterior (MGUH 29466). Locality A-1230.
12: Same specimen showing part of the ventral interior (MGUH 29466).

13–18. **Strophomenoidea gen. et sp. indet. 4**
13: Fragment showing ventral exterior. Note the characteristic ornament (MGUH 29467). Locality A-1230.
14: Fragment of dorsal valve showing the exterior with ornament (MGUH 29468). Locality A-1230.
15, 16: Fragmented shells showing hingeline. Note also the tooth in 16 (MGUH 29469; MGUH 29470). Locality A-1230.
17: Fragment showing dorsal cardinalia (MGUH 29471). Locality A-1230.
18: Close-up of possible alae on a ventral valve showing ornament (MGUH 29472). Locality A-1230.

19–24. **Sowerbyella (Sowerbyella) rectangularis n. sp.**
19: View of ventral valve exterior. Note the almost rectangular outline and the closely spaced costellate ornament (MGUH 29473, holotype). Locality A-1230.
20: View of dorsal valve exterior (MGUH 29474). Locality A-1230.
21: Posterior view of conjoined specimen showing interareas, delthyrial cavity and a short ventral median septum. Ventral valve below (MGUH 29473, holotype).
22, 23: Perpendicular and anterior views of ventral interior. Note the short ventral median septum on fig. 23 (MGUH 29475). Locality A-1230.
24: View of dorsal valve interior. Note the lack of a dorsal median septum (MGUH 29476). Locality A-1230.

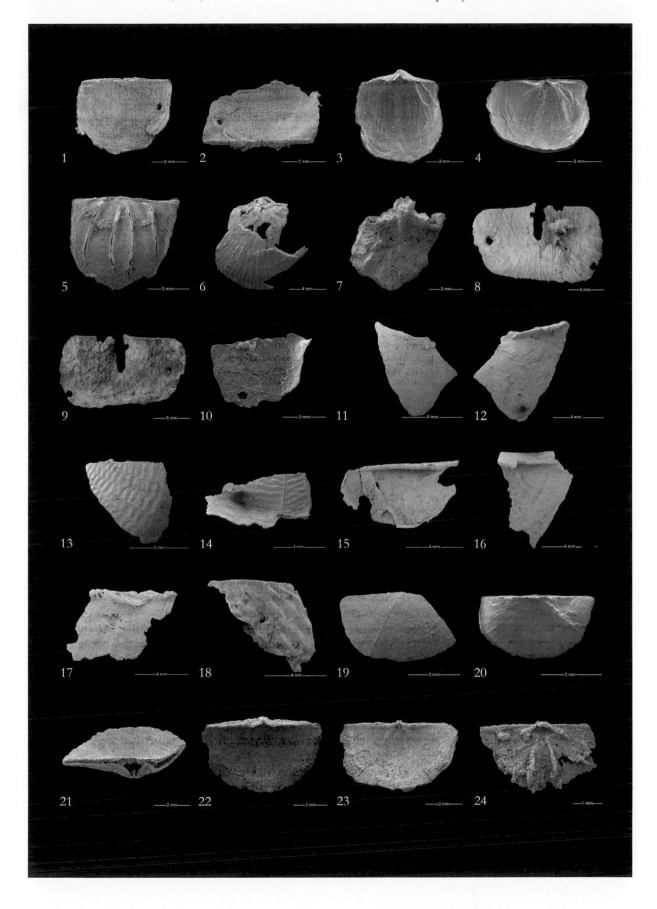

Plate 3

1–5.   *Sowerbyella* (***Rugosowerbyella***) ***praecursor*** **n. sp.**
1:    View of the exterior of a ventral valve. Note the rugate ornament (MGUH 29477). Locality A-1230.
2:    View of dorsal valve exterior (MGUH 29478, holotype). Locality A-1230.
3:    Posterior view of conjoined specimen showing interareas, ventral valve below (MGUH 29479). Locality A-1230.
4:    Interior view of ventral valve. Note the short ventral median septum (MGUH 29477).
5:    Interior view of dorsal valve. Note the two pairs of side septa developed (MGUH 29478, holotype).

6, 7.   *Sowerbyella* (***Rugosowerbyella***) **sp. 1**
6, 7:   Ventral valve exterior and interior views (MGUH 29480). Locality A-1230.

8–12.   *Anisopleurella tricostata* **n. sp.**
8:    Exterior view of ventral valve (MGUH 29481, holotype). Locality A-1230.
9:    Exterior view of dorsal valve (MGUH 29482). Locality A-1230.
10, 11:   perpendicular and anterior view of ventral interior (MGUH 29483). Locality A-1230.
12:    View of dorsal valve showing the characteristic interior (MGUH 29484). Locality A-1230.

13–16.   *Eoplectodonta* (***Eoplectodonta***) **sp.**
13, 14:   Dorsal valve showing the exterior ornament and interior structures (MGUH 29485). Note the low, delayed median septum in figure 14. Locality A-1230.
15:    Dorsal fragment showing exterior ornament and the most posterior parts of the interior structures (MGUH 29486). Locality A-1230.

16–19.   *Ptychoglyptus pauciradiatus* **Reed, 1932**
16:    Exterior view of ventral valve (MGUH 29487). Locality A-1230.
17:    Exterior view of dorsal fragment showing ornament (MGUH 29488). Locality A-1230.
18:    Interior view of ventral valve (MGUH 29487).
19:    Interior view of dorsal fragment, showing chilidium and septae (MGUH 29489). Locality A-1230.

20–24.   *Ptychoglyptus alaensis* **n. sp.**
20:    View of ventral valve exterior showing the alate cardinal extremities (MGUH 29490, holotype). Locality A-1230.
21:    View of dorsal valve exterior (MGUH 29491). Locality A-1230.
22:    Interior view of the holotype (MGUH 29490).
23:    View showing dorsal valve interior (same specimen as in figure 21).
24:    close-up of dorsal cardinalia (MGUH 29492). Locality A-1230.

Plate 4

1, 2.   *Sowerbyites* **sp.**
1, 2:   Dorsal exterior and interior views. Note the lamellose anterior commissure (MGUH 29493). Locality A-1230.

3–10.   *Bimuria* (*Bimuria*) *gilbertella* **Potter, 1991**
3:   View of ventral exterior (MGUH 29494). Locality A-1230.
4:   View of dorsal exterior (MGUH 29495). Locality A-1230.
5:   Lateral view of the same specimen showing dorsal exterior median septum.
6:   Dorsal view of conjoined specimen showing interareas (MGUH 29496). Locality A-1230.
7:   Interior view of ventral valve. Note foramen (MGUH 29494).
8, 9:   Perpendicular and posterior views of dorsal interior. Note the short median septum (MGUH 29497). Locality A-1230.
10:   Close-up of ventral delthyrium showing interarea, teeth and apical foramen (MGUH 29498). Locality A-1230.

11–17.   *Craspedelia potterella* **n. sp.**
11:   Exterior view of ventral valve. Note the sharp geniculation along the commissure (MGUH 29499, holotype). Locality A-1230.
12:   Exterior view of dorsal valve (MGUH 29500). Locality A-1230.
13:   Posterior view of the same specimen showing interarea and notothyrium.
14, 15:   Perpendicular and anterior views of ventral interior (MGUH 29501). Locality A-1230.
16, 17:   Perpendicular and posterior views of dorsal interior (MGUH 29502). Locality A-1230.

18–24.   *Leptellina* (*Leptellina*) *occidentalis* **Ulrich & Cooper, 1936**
18–20:   Views of ventral exterior, interior and close up of teeth and deltidium (MGUH 29503). Locality A-1230.
21, 22:   Dorsal exterior and interior views of morphotype with well-developed subperipheral rim (MGUH 29504). Locality A-1230.
23:   Close-up of cardinalia of same specimen.
24:   Dorsal exterior of a morphotype almost lacking subperipheral rim (MGUH 29505). Locality A-1230.

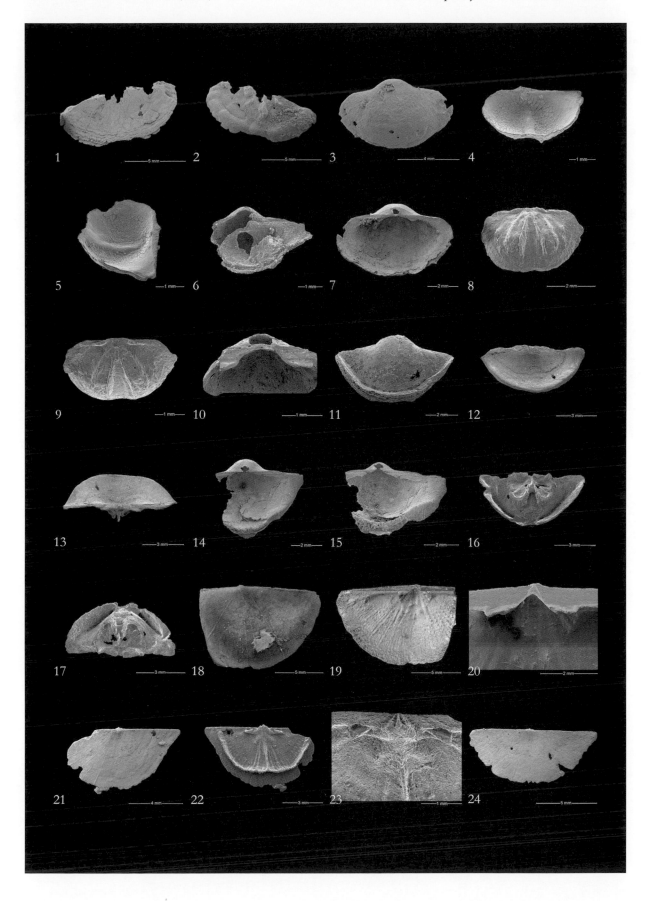

Plate 5

1, 2.   *Leptellina* (*Leptellina*) *occidentalis* Ulrich & Cooper, 1936 (cont.)
1.   Dorsal interior of a morphotype almost lacking subperipheral rim (MGUH 29505).
2.   Close-up of cardinalia of same specimen.

3–8.   *Leptellina* (*Leptellina*) *tennesseensis* Ulrich & Cooper, 1936
3:     Exterior view of fragmented ventral valve (MGUH 29506). Locality A-1230.
4:     Exterior view of fragmented dorsal valve (MGUH 29507). Locality A-1230.
5:     Fragment showing ventral interior (MGUH 29508). Locality A-1230.
6:     Interior view of dorsal valve. Note the large platform (MGUH 29507).
7, 8:   Perpendicular and posterior views of dorsal cardinalia (MGUH 29509). Locality A-1230.

9–14.   *Leptelloidea leptelloides* Bekker, 1922
9:     Exterior view of ventral fragment (MGUH 29510). Locality A-1230.
10:     Posterior view of ventral valve showing interarea and teeth (MGUH 29511). Locality A-1230.
11, 12:   Exterior and interior views of dorsal valve. Note difference in configuration of platform compared to *Leptellina* (MGUH 29512). Locality A-1230.
13, 14:   Close-up of dorsal cardinalia showing 'wing-like' brachiophores (MGUH 29513). Locality A-1230.

15–20.   *Diambonia discuneata* Lamont, 1935
15.   Exterior view of ventral fragment (MGUH 29514). Locality 79WG19.
16.   Ventral exterior view of conjoined specimen (MGUH 29515). Locality 79WG126.
17.   Dorsal exterior view of the same specimen.
18.   Posterior view of ventral valve (MGUH 29514).
19.   Interior view of ventral valve (MGUH 29516). Locality 79WG19.
20.   Interior view of ventral valve. Note the anteriorly elevated median septum and the transverse ridge confining the muscle field (MGUH 29517). Locality 79WG19.

21–24.   *Xenambonites revelatus* Williams, 1962
21:     Exterior view of ventral valve (MGUH 29518). Locality A-1230.
22:     Exterior view of dorsal valve. Note the median depression (MGUH 29519). Locality A-1230.
23:     Interior view of ventral valve (MGUH 29520). Locality A-1230.
24:     Posterior view of same specimen, showing interarea and teeth.

Plate 6

1.   *Xenambonites revelatus* **Williams, 1962** (cont.)
1:       Interior view of dorsal valve (MGUH 29521). Locality A-1230.

2–7.   **New genus aff.** *Aegiria*
2:       Fragment showing dorsal exterior (MGUH 29522). Locality 79WG19.
3:       Fragment showing ventral delthyrium and interarea (MGUH 29523). Locality 79WG19.
4, 5:   Different fragments of dorsal interior showing platform rim (MGUH 29522, MGUH 29524). Locality 79WG19.
6, 7:   Different fragments of dorsal interior showing cardinalia and bema-like structure (MGUH 29525, MGUH 29526). Locality 79WG19.

8–14.   *Anoptambonites grayae* **(Davidson, 1883)**
8:       Exterior view of ventral valve (MGUH 29527). Locality A-1230.
9:       Exterior view of dorsal valve (MGUH 29528). Locality A-1230.
10, 11: Posterior and perpendicular views of ventral interior showing interarea with teeth. Note the short ventral median septum in 11 (MGUH 29529). Locality A-1230.
12, 13: Perpendicular and oblique views of dorsal interior Note the extremely high dorsal median septum (MGUH 29530). Locality A-1230.
14:      Close-up of valve exterior showing punctae (MGUH 29527).

15–20. *Anoptambonites pulchra* **(Cooper, 1956)**
15:      Exterior view of ventral valve (MGUH 29531). Locality A-1230.
16:      Exterior view of dorsal valve (MGUH 29532). Locality A-1230.
17, 18: Ventral and dorsal exterior views of conjoined specimen showing interareas, dorsal median septum and possible geniculation. Ornamentation is covered by glue (MGUH 29533). Locality A-1230.
19:      Interior view of ventral valve (MGUH 29534). Locality A-1230.
20:      Interior view of dorsal valve (MGUH 29535). Locality A-1230.

21–22. *Anoptambonites* **sp.**
21, 22: Exterior and interior views of dorsal valve (MGUH 29536). Locality 79WG126.

23–24. *Kassinella* (*Trimurellina?*) **sp.**
23:      Exterior view of ventral valve (MGUH 29537). Locality 79WG19.
24:      Exterior view of dorsal fragment (MGUH 29538). Locality 79WG19.

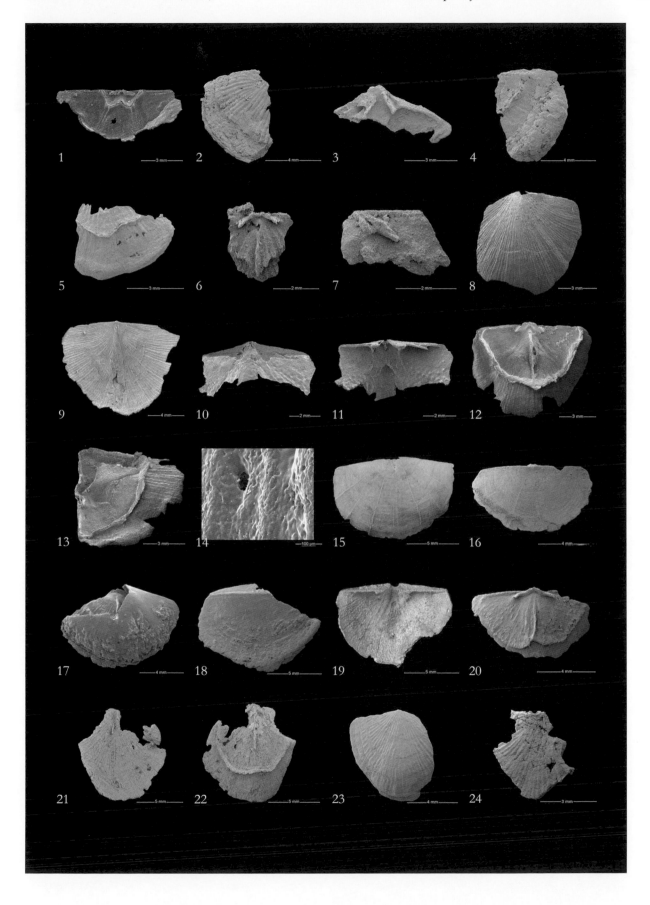

Plate 7

1–5.   *Kassinella* (*Trimurellina?*) **sp.** (cont.)
1, 2:   Interior views of ventral valve (MGUH 29537, MGUH 29539). Locality 79WG19.
3–5:   Perpendicular, anterior and oblique views of dorsal interior. Note both platform and subperipheral rim is visible in 4 and 5. Further, note the position of the dorsal median septum (MGUH 29540, MGUH 29538). Locality 79WG19.

6, 7.   *Sampo?* **sp.**
6, 7:   Ventral exterior and interior views (MGUH 29541). Locality 79WG126.

8, 9.   **Plectambonitoidea gen.** *et* **sp. indet.**
1 8, 9: Fragment showing ventral exterior and interior (MGUH 29542). Locality 79WG126.

10, 11. *Atelelasma?* **sp.**
10:   Fragment of ventral? valve (MGUH 29543). Locality A-1230.
11:   Close-up of ornament on same specimen.

12, 13. *Triplesia* **sp.**
12:   Exterior view of dorsal valve (MGUH 29544). Locality A-1230.
13:   Interior view of same specimen.

14.   *Grammoplecia?* **sp.**
14:   Fragment of valve exterior showing characteristic ornament (MGUH 29545). Locality 79WG19.

15–20.   *Skenidioides multifarius* **Potter, 1990b**
15:   Exterior view of ventral valve (MGUH 29546). Locality A-1230.
16:   Exterior view of dorsal valve (MGUH 29547). Locality A-1230.
17:   Posterior view of conjoined specimen, ventral valve below. Note the deep delthyrium (MGUH 29548). Locality A-1230.
18, 19: Perpendicular, anterior view of ventral interior showing spondylium simplex (MGUH 29549). Locality A-1230.
20:   Interior view of dorsal valve. Note that the cardinal process extends anteriorly as a median septum (MGUH 29550). Locality A-1230.

21–24.   *Replicoskenidioides* **sp.**
21:   Ventral exterior view of conjoined specimen. Note the cardinal process and dorsal median septum is visible through the hole in the ventral valve (MGUH 29551). Locality 79WG19.
22:   Dorsal exterior view of the same specimen.
23:   Posterior view of the same specimen. Note the cardinal process and dorsal median septum.
24:   Oblique view of dorsal interior showing median septum and brachiophores (MGUH 29552). Locality 79WG19.

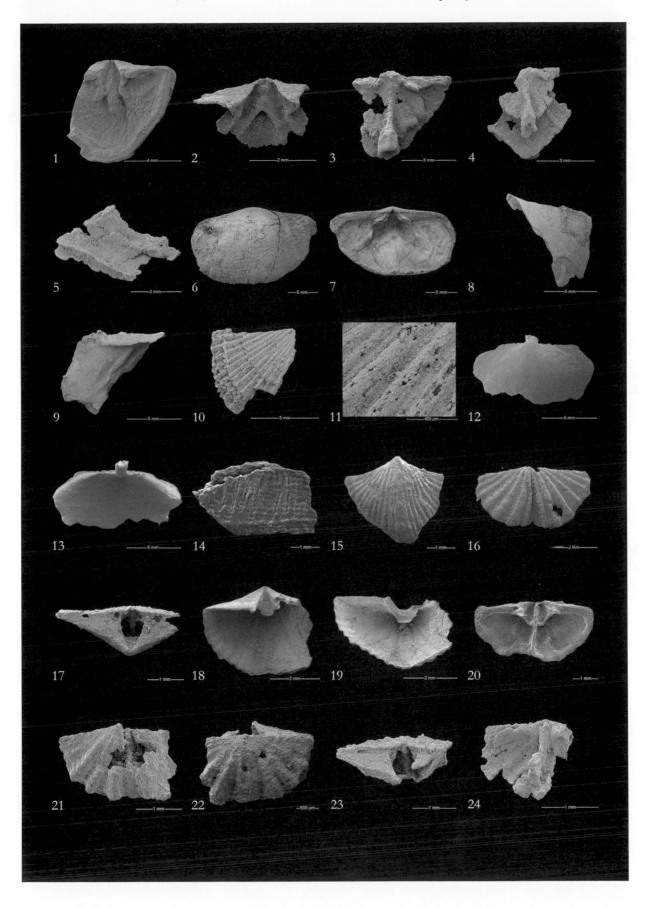

Plate 8

1, 2.   Skenidiid n. gen. sp. indet.
1, 2:   Exterior and interior views of dorsal valve (MGUH 29553). Locality A-1230.

3–6.   *Plectorthis* sp.
1 3:   Fragment showing ventral exterior (MGUH 29554). Locality A-1230.
4:   Fragment showing dorsal exterior (MGUH 29555). Locality A-1230.
5:   Fragment showing ventral callus and diductor muscle impressions (MGUH 29556). Locality A-1230.
6:   Interior view of dorsal valve (MGUH 29557). Locality A-1230.

7–10.   *Plectorthis* sp. 2
7:   Exterior view of ventral valve (MGUH 29558). Locality A-1230.
8:   Exterior view of dorsal valve (MGUH 29559). Locality A-1230.
9:   Interior view of ventral valve (MGUH 29558).
10:   Interior view of dorsal valve. Note the blade-like cardinal process (MGUH 29559).

11, 12.   *Desmorthis*? sp.
11:   Exterior view of dorsal valve (MGUH 29560). Locality A-1230.
12:   Interior view of dorsal valve. Note the blade-like cardinal process (MGUH 29560).

13–17.   *Doleroides* sp.
13:   Exterior view of ventral valve (MGUH 29561). Locality A-1230.
14:   Exterior view of dorsal valve (MGUH 29562). Locality A-1230.
15:   Interior view of ventral valve (MGUH 29561).
16:   Interior view of dorsal valve (MGUH 29562).
17:   Close-up of ornament (MGUH 29561).

18–20.   *Doleroides* n. sp. aff. *D. panna* (Andreeva)
18:   Exterior view of ventral valve (MGUH 29563). Locality A-1230.
19, 20:   Perpendicular and anterior views of the interiors of the same specimen.

21–24.   *Gelidorthis perisiberiaensis* n. sp.
21:   Exterior view of ventral valve (MGUH 29564). Locality A-1230.
22:   Exterior view of dorsal valve (MGUH 29565). Locality A-1230.
23, 24:   Perpendicular and anterior views of ventral interior (MGUH 29566, holotype). Locality A-1230.

Plate 9

1.   ***Gelidorthis perisiberiaensis* n. sp.** (cont.)
1:   Interior view of dorsal valve. Note that the cardinal process is 'swollen' posteriorly (MGUH 29567). Locality A-1230.

2–4.   **Cyclocoeliididae gen.** *et* **sp. indet.**
2, 3: Exterior and interior views of ventral valve. Note the teeth (MGUH 29568). Locality A-1230.
4:   Anterior view of ventral valve showing the interior of the same specimen.

5–9.   ***Phragmorthis buttsi* Cooper, 1956**
5:   Exterior view of ventral valve (MGUH 29569). Locality A-1230.
6:   Exterior view of dorsal valve (MGUH 29570). Locality A-1230.
7:   Interior view of ventral valve. Note the deep delthyrium (MGUH 29571). Locality A-1230.
8:   Interior view of dorsal valve. Note the absence of a cardinal process (MGUH 29572). Locality A-1230.
9:   Oblique view of same specimen showing shape of median septum (MGUH 29572).

10–14. ***Rhactorthis* sp.**
10:   Exterior view of ventral valve (MGUH 29573). Locality A-1230.
11:   Interior view of ventral valve (MGUH 29573).
12:   Exterior view of dorsal valve (MGUH 29574). Locality A-1230.
13, 14: Dorsal interior views of different specimens (MGUH 29574, MGUH 29575). Locality A-1230.

15–22. ***Scaphorthis virginiensis* Cooper, 1956**
15:   Exterior view of ventral valve (MGUH 29576). Locality A-1230.
16:   Exterior view of dorsal valve (MGUH 29577). Locality A-1230.
17:   Posterior view of conjoined specimen. Dorsal valve above (MGUH 29578). Locality A-1230.
18, 19: Perpendicular and anterior views of ventral interior (MGUH 29576).
20:   Interior view of dorsal valve. Cardinal process broken off (MGUH 29579).
21:   Interior view of the posterior part of a conjoined specimen. Dorsal valve above, note the blade-like cardinal process developed in this specimen (MGUH 29580).
22:   Close-up of impunctate ornament (MGUH 29578).

23–24. ***Diochtofera* aff. *conspicua* Potter, 1990a**
23:   Fragment of dorsal valve showing exterior ornament (MGUH 29581).
24:   Fragment of dorsal valve showing part of the interior (same specimen as above).

Plate 10

1–2.   *Diochtofera* aff. *conspicua* **Potter, 1990a** (cont.)
1, 2:   Different views of dorsal interior (MGUH 29582, MGUH 29583). Locality 79WG19.

3–11.   *Taphrorthis immatura* **Williams & Curry, 1985**
3:   Exterior view of ventral valve (MGUH 29584). Locality A-1230.
4:   Ventral valve interior showing long callus and bilobed muscle field (MGUH 29584).
5:   Fragment of ventral interior showing long callus and bilobed muscle field (MGUH 29585). Locality A-1230.
6:   Exterior view of dorsal valve in a probable juvenile specimen (MGUH 29586). Locality A-1230.
7:   Interior view of same specimen. Note the almost sessile notothyrium compared to figure 9 and 10.
8:   Exterior view of dorsal valve in an adult specimen (MGUH 29587). Locality A-1230.
9, 10:   Perpendicular and anterior views of dorsal interior in the adult specimen. Note the elevated notothyrial platform and the widely divergent brachiophores (MGUH 29587).
11:   Close-up of exterior ornament (MGUH 29587).

12–15.   *Austinella* **sp.**
12:   Fragment of ventral valve showing ornamentation (MGUH 29588). Locality A-1230.
13:   Fragment of dorsal valve showing the exterior (MGUH 29589). Locality A-1230.
14:   Fragment showing ventral interior (MGUH 29588).
15:   Fragment showing dorsal interior (MGUH 29589).

16–21.   *Dinorthis* **sp.**
16:   Exterior view of ventral valve (MGUH 29590). Locality A-1230.
17:   Exterior view of dorsal valve (MGUH 29591). Locality A-1230.
18:   Interior view of ventral valve (MGUH X29590).
19:   Interior view of dorsal valve (MGUH 29591).
20, 21:   Close-up of ornament showing impunctate nature of valves (MGUH 29590).

22–24.   *Ptychopleurella uniplicata* **Cooper, 1956**
22:   Exterior view of ventral valve (MGUH 29592). Locality A-1230.
23:   Exterior view of dorsal valve (MGUH 29593). Locality A-1230.
24:   Posterior view of conjoined specimen. Ventral valve below (MGUH 29594). Locality A-1230.

Plate 11

1–3.   ***Ptychopleurella uniplicata* Cooper, 1956**
1, 2:   Perpendicular and anterior views of ventral valve (MGUH 29595; MGUH 29596). Locality A-1230.
3:   Interior view of dorsal valve (MGUH 29597). Locality A-1230.

4–9.   ***Glyptorthis* sp.**
4:   Exterior view of ventral valve (MGUH 29598). Locality 79WG126.
5:   Interior view of ventral valve (MGUH 29598). Locality 79WG126.
6:   Exterior view of ventral valve (MGUH 29599). Locality A-1230.
7:   Exterior view of dorsal valve (MGUH 29600). Locality A-1230.
8:   Interior view of ventral valve (MGUH 29601). Locality A-1230.
9:   Interior view of dorsal valve (MGUH 29600).

10–12.   ***Hesperorthis* sp.**
10:   Exterior view of ventral valve (MGUH 29602). Locality 79WG126.
11:   Perpendicular interior view of same specimen showing ventral interarea and teeth.
12:   Oblique anterior view of same specimen showing ventral interior.

13–17.   ***Hesperorthis* sp. 2**
13:   Exterior view of ventral valve (MGUH 29603). Locality A-1230.
14:   Exterior view of dorsal valve (MGUH 29604). Locality A-1230.
15:   Interior view of ventral valve (MGUH 29603).
16:   Interior view of dorsal valve (MGUH 29605). Locality A-1230.
17:   Interior view of dorsal valve (MGUH 29606). Locality A-1230.

18–24.   ***Duolobella sandiae* n. gen. *et* n. sp.**
18, 19, 20: Perpendicular, anterior and posterior views of ventral exterior (MGUH 29607, holotype). Locality A-1230.
21, 22:   Perpendicular and anterior views of dorsal exterior (MGUH 29608). Locality A-1230.
23, 24:   Perpendicular and anterior views of ventral interior (MGUH 29607).

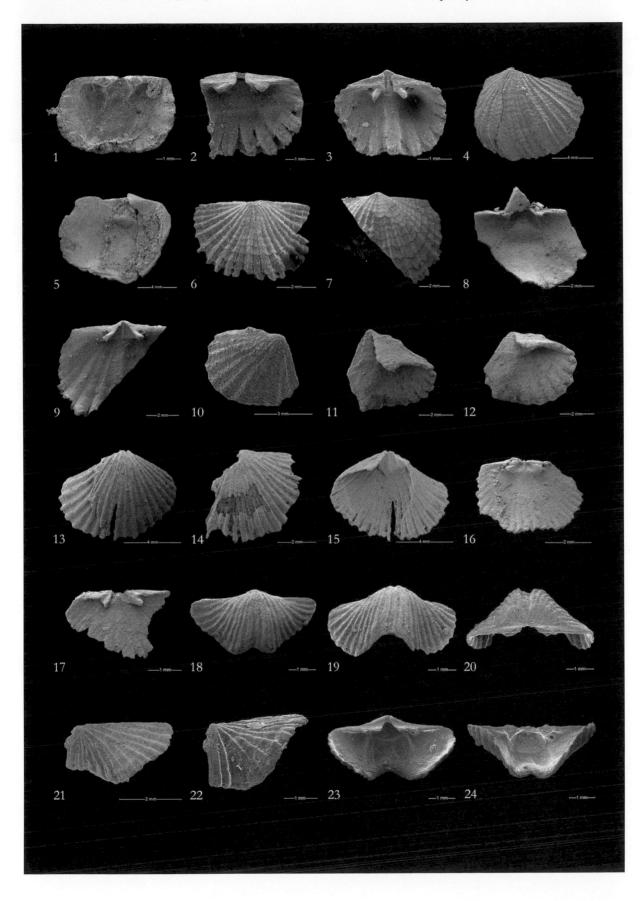

Plate 12

1–2.   ***Duolobella sandiae*** **n. gen.** *et* **n. sp.** (cont.)
1, 2:   Interior view of dorsal valve. Note the weakly developed blade-like cardinal process (MGUH 29609, MGUH 29610). Locality A-1230.

3–8.   ***Palaeowingella farewellensis*** **n. gen.** *et* **n. sp.**
3:   Exterior view of ventral valve (MGUH 29611, holotype). Locality A-1230.
4:   Exterior view of dorsal valve (MGUH 29612). Locality A-1230.
5:   Posterior view of ventral exterior (MGUH 29611, holotype).
6, 7:   Perpendicular and anterior views of ventral interior. Note that the spondylium is not supported by a median septum (MGUH 29611, holotype).
8:   Interior view of dorsal valve (MGUH 29612).

9–11.   **Orthoidea gen.** *et* **sp. indet. 1**
9:   Exterior view of ventral valve (MGUH 29613). Locality A-1230.
10:   Anterior view of ventral exterior (MGUH 29613).
11:   Interior view of ventral valve (MGUH 29613).

12, 13.   **Orthidina gen.** *et* **sp. indet.**
1 12, 13:   Exterior and interior view ventral valve (MGUH 29614). Locality A-1230.

14–17.   **Orthidina gen.** *et* **sp. indet.**
2 14:   Exterior view of ventral valve (MGUH 29615). Locality A-1230.
15:   Exterior view of dorsal valve (MGUH 29616). Locality A-1230.
16:   Interior view of ventral valve (MGUH 29615).
17:   Interior view of dorsal valve (MGUH 29616).

18, 19.   **Orthidina gen.** *et* **sp. indet. 3**
18, 19:   Exterior and interior views of dorsal valve (MGUH 29617). Locality A-1230.

20–21.   **Orthidina gen.** *et* **sp. indet.**
4 20, 21:   Perpendicular and posterior views of ventral exterior (MGUH 29618). Locality A-1230.
22:   Interior view of the same specimen.

23–24.   **Orthidina gen.** *et* **sp. indet. 5**
23:   Exterior view of ventral fragment (MGUH 29619). Locality A-1230.
24:   Interior view of same specimen (MGUH 29619).

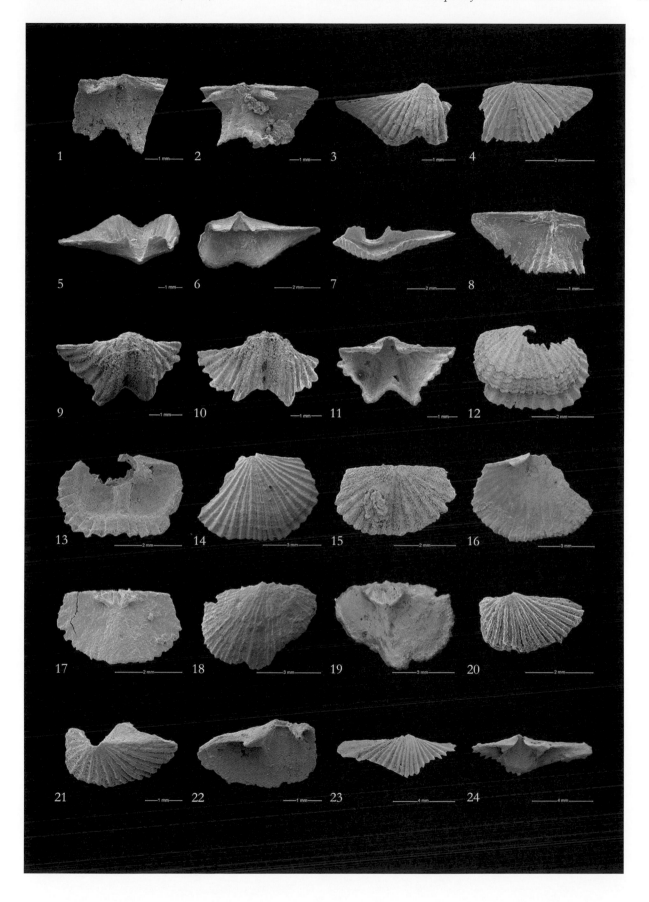

Plate 13

1–5.   Orthidina gen. *et* sp. indet. 6
1:      Exterior view of ventral valve (MGUH 29620). Locality A-1230.
2:      Exterior view of dorsal valve (MGUH 29621). Locality A-1230.
3:      Interior view of ventral valve (MGUH 29620).
4:      Interior view of dorsal valve (MGUH 29621).
5:      Close-up of exterior ornament showing possible borings (MGUH 29620).

6–9.   *Paucicrura?* sp.
6:      Exterior view of ventral valve (MGUH 29622). Locality A-1230.
7:      Interior view of ventral valve (MGUH 29623). Locality A-1230.
8, 9:   Exterior and interior views of dorsal valve. Note the ring of muscle bounding ridges (MGUH 29624). Locality A-1230.

10–12.   *Dicoelosia jonesridgensis* Ross & Dutro, 1966
10:     Ventral view of the exterior of a conjoined specimen (MGUH 29625). Locality 79WG19.
11:     Dorsal exterior view of the same specimen.
12:     Posterior view of the same specimen, ventral valve below.

13–15.   *Epitomyonia relicina* Potter, 1990b
13, 14: Dorsal and posterior views of conjoined specimen. Ventral valve below in 14 (MGUH 29626). Locality 79WG19.
15:     Anterior view of same specimen showing dorsal interior. Ventral valve below.

16–21.   *Diorthelasma parvum* Cooper, 1956
16:     Exterior view of ventral valve (MGUH 29627). Locality A-1230.
17:     Exterior view of dorsal valve (MGUH 29628). Locality A-1230.
18:     Interior view of ventral valve (MGUH 29629). Locality A-1230.
19–21: Perpendicular and anterior views of dorsal interior (MGUH 29630, MGUH 29631). Locality A-1230.

22–24.   *Oanduporella kuskokwimensis* n. sp.
22, 23: Perpendicular and oblique views of ventral exterior (MGUH 29632). Locality A-1230.
24:     Exterior view of dorsal valve (MGUH 29633, holotype). Locality A-1230.

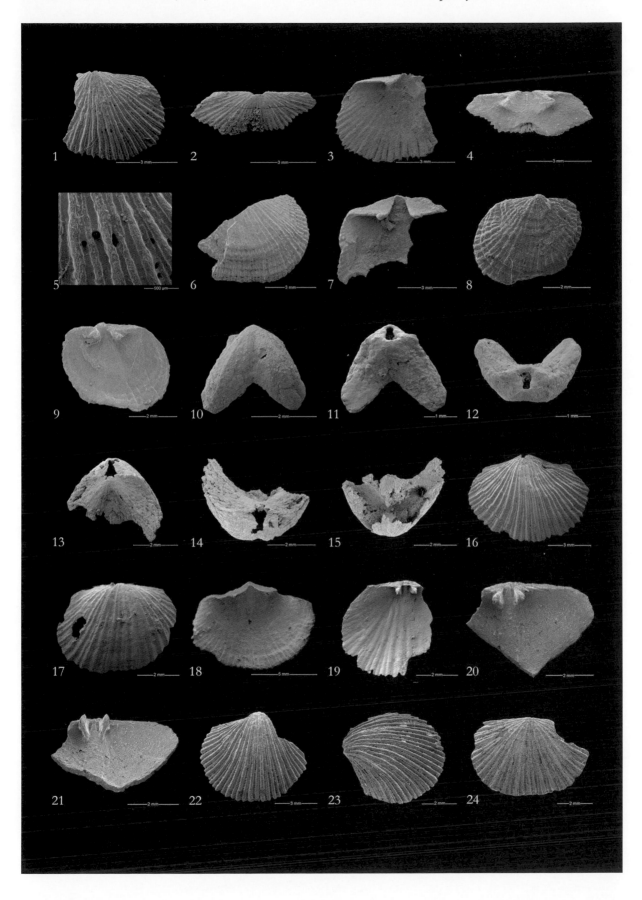

Plate 14

1–6.   *Oanduporella kuskokwimensis* **sp. nov.** (cont.)
1, 2:   Perpendicular and anterior views of ventral interior (MGUH 29634). Locality A-1230.
3, 4:   Perpendicular and anterior views of dorsal interior (MGUH 29633, holotype).
5, 6:   Close-up of punctate exterior ornament (MGUH 29632).

7, 8.   *Salopina?* **sp.**
7:   Ventral view of conjoined specimen (MGUH 29635). Locality 79WG19.
8:   Dorsal view of same specimen.

9–13.   *Callositella cheeneetnukensis* **n. gen.** *et* **n. sp.**
9:   Exterior view of ventral valve (MGUH 29636). Locality A-1230.
10:   Exterior view of dorsal valve (MGUH 29637, holotype). Locality A-1230.
11:   Interior view of ventral valve showing strongly impressed pinnate mantle canal system (MGUH 29636).
12, 13:   Interior view of dorsal valve. Note the pair of callosities bounding the anterior margin of the median septum (MGUH 29637, holotype, MGUH 29638). Locality A-1230.

14.   *Laticrura* **aff.** *pionodema* **Cooper, 1956**
14:   Exterior view of ventral valve (MGUH 29639). Locality A-1230.
15:   Exterior view of dorsal valve (MGUH 29640). Locality A-1230.
16:   Interior views of ventral valve (MGUH 29641). Locality A-1230.
17:   Interior view of dorsal valve (MGUH 29642). Locality A-1230.

18, 19.   *Camerella* **sp.**
18, 19:   Exterior and interior views of dorsal valve (MGUH 29643). Locality A-1230.

20–23.   *Brevicamera?* **sp.**
20:   Exterior view of ventral valve (MGUH 29644). Locality 79WG19.
21, 22:   Perpendicular and lateral views of dorsal exterior (MGUH 29645). Locality 79WG19.
23:   Interior view of ventral valve (MGUH 29646). Locality 79WG19.

24.   *Perimecocoelia semicostata* **Cooper, 1956**
24:   Exterior view of ventral valve (MGUH 29647). Locality A-1230.

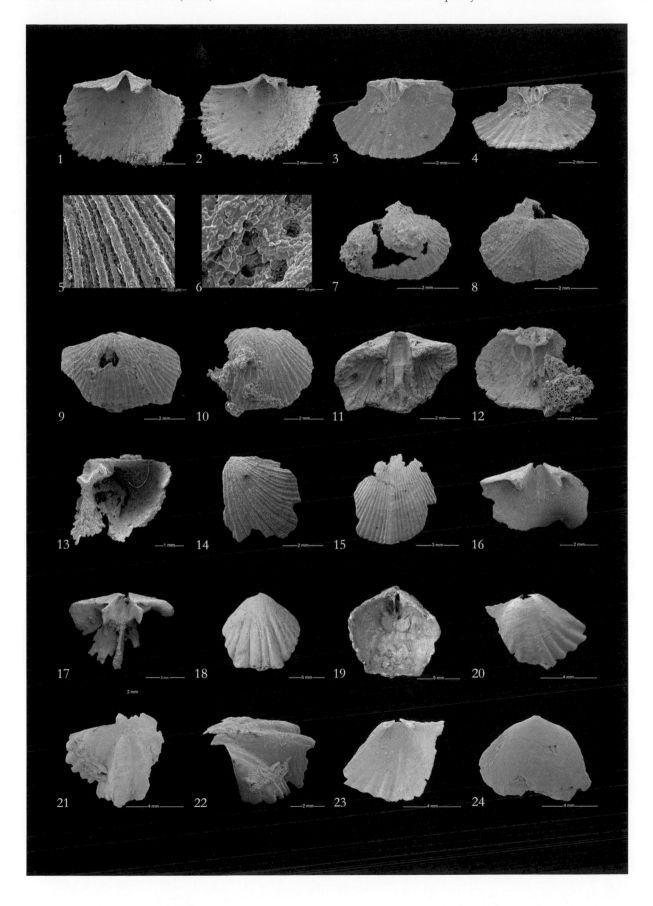

Plate 15

1–3.   *Perimecocoelia semicostata* Cooper, 1956 (cont.)
1:      Exterior view of dorsal valve (MGUH 29648). Locality A-1230.
2:      Interior view of ventral valve (MGUH 29647).
3:      Interior view of dorsal valve (MGUH 29648).

4–7.   *Stenocamara*? sp.
4:      Exterior view of ventral valve (MGUH 29649). Locality 79WG19.
5:      Exterior view of dorsal valve (MGUH 29650). Locality 79WG19.
6:      Interior view of ventral valve (MGUH 29649).
7:      Interior view of dorsal valve (MGUH 29650).

8–13.   *Eoanastrophia*? sp.
8:      Exterior view of ventral valve (MGUH 29651). Locality 79WG19.
9:      Exterior view of dorsal valve (MGUH 29652). Locality 79WG19.
10:     Interior view of ventral valve (MGUH 29651).
11:     Oblique lateral view of ventral interior. Note that the short ventral median septum does not support the spondylium for entire length (MGUH 29651).
12:     Interior view of dorsal valve (MGUH 29652).
13:     Oblique lateral view of dorsal interior showing inner and outer plates (MGUH 29652).

14, 15.   *Didymelasma*? sp.
14, 15:   Exterior and interior views ventral valve (MGUH 29653). Locality A-1230.

16, 17.   *Schizostrophina*? sp.
16, 17:   Exterior and interior views ventral valve (MGUH 29654). Locality 79WG19.

18–20.   *Brevilamnulella* sp.
18:     Exterior view of dorsal valve (MGUH 29655). Locality 79WG19.
19:     Interior view of same specimen.
20:     Oblique anterior view of dorsal interior showing the configuration of inner and outer plates.

21–24.   *Galeatellina* n. sp.
21:     Exterior view of ventral valve (MGUH 29656). Locality 79WG126.
22:     Exterior view of dorsal valve (MGUH 29657). Locality 79WG19.
23:     Lateral view of ventral exterior showing palintrope (MGUH 29656).
24:     Lateral view of dorsal exterior (MGUH 29657).

Plate 16

1–3.   *Galeatellina* sp. nov.
1:   Interior view of ventral valve (MGUH 29656).
2:   Oblique anterior view of ventral interior (MGUH 29656).
3:   Interior view of dorsal valve (MGUH 29657).

4–7.   *Orthorhynchuloides*? sp.
4:   Exterior view of ventral valve (MGUH 29658). Locality 79WG19.
5:   Fragment showing the exterior of a dorsal valve (MGUH 29659). Locality 79WG19.
6:   Exterior view of dorsal valve (MGUH 29660). Locality 79WG19.
7:   Interior view of dorsal valve (MGUH 29660).

8–10.   *Anazyga* sp.
8:   Ventral exterior view of a conjoined specimen (MGUH 29661). Locality 79WG126.
9:   Dorsal exterior view of the same specimen.
10:   Posterior view of the same specimen.

11.   *Zygospira*? sp.
11:   Exterior view of ventral valve (MGUH 29662). Locality 79WG19.

12.   *Catazyga*? sp.
12:   Exterior view of ventral valve (MGUH 29663). Locality 79WG126.

13–19.   *Cyclospira orbus* Cocks and Modzalevskaya, 1997
13:   Ventral exterior view of conjoined specimen. Note spiralia and dorsal median septum is visible through the hole in the ventral valve (MGUH 29664). Locality 79WG19.
14:   Dorsal exterior view of conjoined specimen (MGUH 29665). Locality 79WG19.
15:   Posterior view of conjoined specimen, ventral valve below (MGUH 29664).
16:   Interior view of ventral valve (MGUH 29666). Locality 79WG19.
17:   Interior view of dorsal valve (MGUH 29667). Locality 79WG19.
18:   Close-up of spiralia on specimen (MGUH 29664).
19:   Close-up of micro punctuated ornament (MGUH 29664).

20–23.   *Cyclospira elegantula* Rozman, 1964
20:   Exterior view of ventral valve (MGUH 29668). Locality A-1230.
21:   Dorsal exterior view of conjoined specimen (MGUH 29669). Locality A-1230.
22:   Posterior view of ventral valve (MGUH 29668).
23:   Oblique view of conjoined specimen (MGUH 29669).

24.   Fam., gen. *et* sp. indet. 1
24:   Exterior view of ventral valve (MGUH 29670). Locality A-1230.

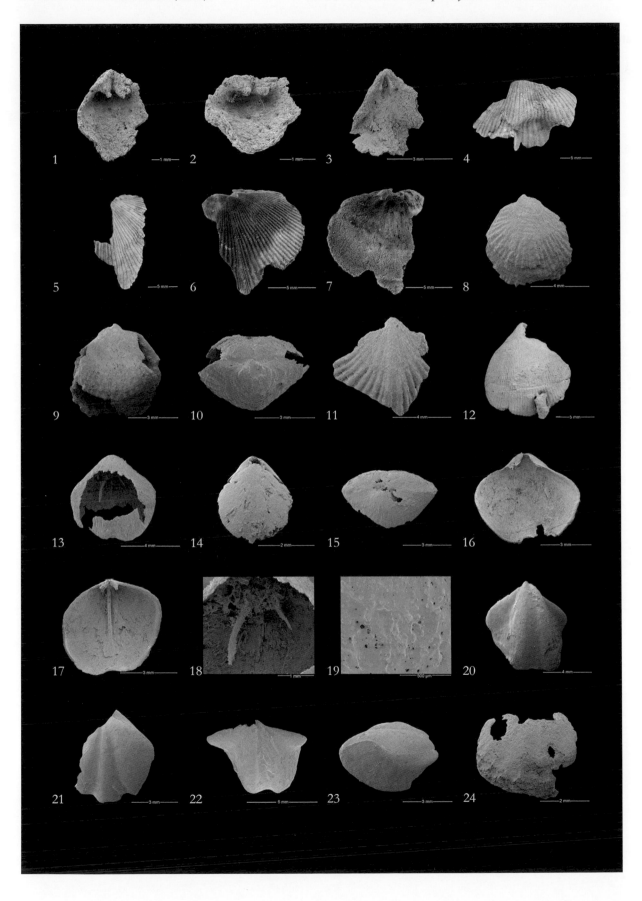

Plate 17

1–3.   **Fam., gen. *et* sp. indet. 1** (cont.)
1:   Interior view of ventral valve (MGUH 29670).
2:   Exterior view of ventral? valve (MGUH 29671). Locality 79WG126.
3:   Interior view of ventral? valve (MGUH 29671).

4–6.   **Fam., gen. *et* sp. indet. 2**
4:   Exterior view of ventral valve (MGUH 29672). Locality 79WG126.
5:   Posterior view of ventral valve (MGUH 29672). 6: Interior view of same specimen.

7, 8.   **Fam., gen. *et* sp. indet. 3**
7:   Exterior view of ventral valve (MGUH 29673). Locality 79WG126.
8:   Interior view of ventral valve (MGUH 29673).

9–11.   **Fam., gen. *et* sp. indet. 4**
9:   Exterior view of ventral valve (MGUH 29674). Locality 79WG19.
10:   lateral view of ventral valve (MGUH 29674).
11:   Interior view of same specimen.

12, 13.   **Fam., gen. *et* sp. indet. 5**
12:   Exterior view of ventral valve (MGUH 29675). Locality 79WG19.
13:   Interior view of same specimen.

14–16.   **Fam., gen. *et* sp. indet. 6**
14:   ventral exterior view of conjoined specimen (MGUH 29676). Locality 79WG126.
15:   Dorsal exterior view of same specimen.
16:   Lateral view of same specimen.

17–20.   **Fam., gen. *et* sp. indet. 7**
17:   Exterior view of ventral valve (MGUH 29677). Locality 79WG19.
18:   Interior view of same specimen.
19:   Exterior view of dorsal valve (MGUH 29678). Locality 79WG19.
20:   Interior view of same specimen.

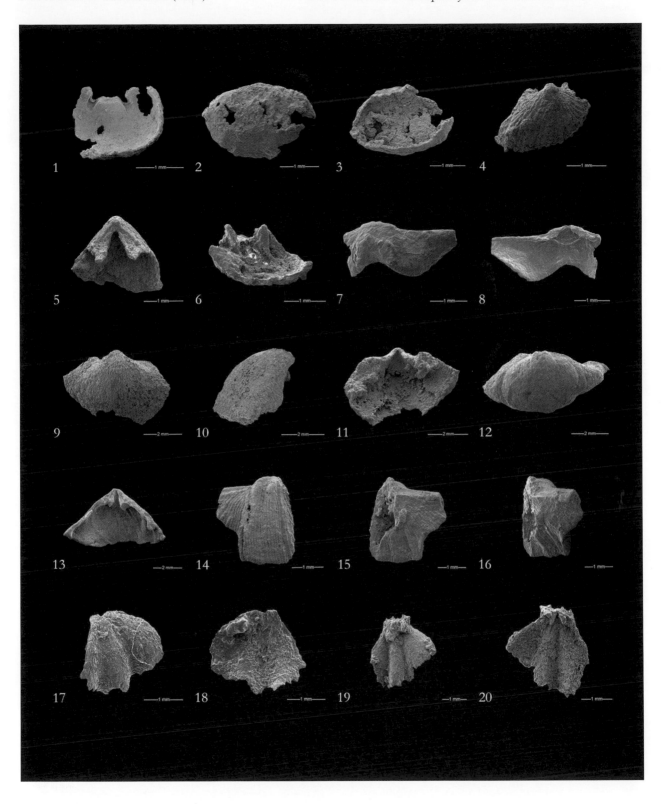